NASA Schedule Management Handbook

NASA STI Program...in Profile

Since its founding, the National Aeronautics and Space Administration (NASA) has been dedicated to the advancement of aeronautics and space science. The NASA Scientific and Technical Information (STI) program plays a key part in helping NASA maintain this important role.

The NASA STI program operates under the auspices of the Agency Chief Information Officer. It collects, organizes, provides for archiving, and disseminates NASA's STI. The NASA STI program provides access to the NASA Aeronautics and Space Database and its public interface, the NASA technical report server, thus providing one of the largest collections of aeronautical and space science STI in the world. Results are published in both non-NASA channels and by NASA in the NASA STI report series, which include the following report types:

- **Technical Publication:** Reports of completed research or a major significant phase of research that presents the results of NASA programs and includes extensive data or theoretical analysis. Includes compilations of significant scientific and technical data and information deemed to be of continuing reference value. NASA counterpart of peer-reviewed formal professional papers but has less stringent limitations on manuscript length and extent of graphic presentations.
- **Technical Memorandum:** Scientific and technical findings that are preliminary or of specialized interest, e.g., quick release reports, working papers, and bibliographies that contain minimal annotation. Does not contain extensive analysis.
- **Contractor Report:** Scientific and technical findings by NASA sponsored contractors and grantees.
- **Conference Publication:** Collected papers from scientific and technical conferences, symposia, seminars, or other meetings sponsored or co-sponsored by NASA.
- **Special Publication:** Scientific, technical, or historical information from NASA programs, projects, and missions, often concerned with subjects having substantial public interest.
- **Technical Translation:** English-language translations of foreign scientific and technical material pertinent to NASA's mission.

Specialized services also include creating custom thesauri, building customized databases, and organizing and publishing research results.

For more information about the NASA STI program, see the following:
- Access the NASA STI program home page at www.sti.nasa.gov
- E-mail your question via the Internet to help@sti.nasa.gov
- Fax your question to the NASA STI help desk at 301-621-0134
- Phone the NASA STI help desk at 301-621-0390
- Write to: NASA STI Help Desk, NASA Center for AeroSpace Information, 7115 Standard Drive, Hanover, MD 21076-1320

NASA/SP-2010-3403

Schedule Management Handbook

National Aeronautics and Space Administration
NASA Headquarters
Washington, D.C. 20546

March 2011

Table of Contents

Figures and Illustrations viii

Preface ix
P.1   Purpose ix
P.2   Applicability ix
P.3   Authority ix
P.4   References x

Acknowledgments xi

Chapter 1   Introduction 1
1.1   Background 1
1.2   Policy Implementation 3

Chapter 2   Schedule Management Overview 4
2.1   Schedule Management Approach 4
2.2   Roles and Responsibilities 8
2.3   Contractor Schedules and Coordination 9
2.4   In-House Schedules and Coordination 10
2.5   External Organizations Schedules and Coordination 13
2.6   Schedule Training 14

Chapter 3   Schedule Management Tool Considerations 16
3.1   Overview 16
3.2   Best Practices 16
3.2.1   Functional Capabilities 16
3.2.2   Interface Capabilities 16
3.2.3   Technical Capabilities 17

Chapter 4   Pre-Schedule Development Activity 18
4.1   Overview 18
4.2   Assignment of Project Planner/Scheduler 18
4.3   Program/Project Scope 18
4.4   Project Work Breakdown Structure (WBS) 19
4.5   Project Organizational Breakdown Structure (OBS) 21
4.6   Project Funding 23
4.7   Project Documentation 24
4.8   Baseline Change Log 24
4.9   Schedule Requirements 25

Chapter 5   Integrated Master Schedule Development 26
5.1   Overview 26
5.2   Data Input and Arrangement 26
5.3   Task/Activity Definition 27

5.4	Task/Activity Sequencing	33
5.5	Duration Estimating	37
5.6	Resource Planning	40
5.7	Schedule Margin Planning	45
5.8	Establishing the IMS Baseline	46

Chapter 6	Status Updates and Schedule Maintenance	51
6.1	Overview	51
6.2	Status Update Accounting	51
6.3	Schedule Maintenance	53
6.4	Schedule Data Back-up and Archive	54

Chapter 7	Schedule Assessment and Analysis	55
7.1	Overview	55
7.2	Levels of Insight	55
7.3	Schedule Logic Credibility Health Check	57
7.4	Critical Path Identification and Analysis	59
7.5	Schedule Performance Trend Analysis	61
7.6	Baseline vs. Current Comparison and Analysis	64
7.7	Schedule Margin Assessment	65
7.8	Validate Cost/Schedule Correlation	66
7.9	Schedule Risk Assessment (SRA)	69
7.10	Duration Compression	71
7.11	Earned Value Schedule Analysis	71

Chapter 8	Schedule Control	75
8.1	Overview	75
8.2	Baseline Content	76
8.3	Baseline Control Process	76
8.4	Re-planning	78
8.5	Re-baselining	78
8.6	Current Schedule Control	78

Chapter 9	Schedule Reporting	80
9.1	Overview	80
9.2	Best Practices	80
9.2.1	Management Summary	81
9.2.2	Schedule Logic Network	81
9.2.3	Critical Path Identification	82
9.2.4	Total Slack Report	82
9.2.5	Schedule Risk	82
9.2.6	Schedule Margin Metrics	83
9.2.7	Performance Trends	83

Chapter 10	Schedule Data and Lessons Learned Archival	84
10.1	Overview	84
10.2	Schedule Archives	84
10.3	Lessons Learned	84

10.4	Historical Narrative	85
10.5	Data Statistics	85

Appendix A – Acronyms	87
Appendix B – Glossary of Terms	90
Appendix C – Data Requirements Descriptions (DRD)	94
Appendix D – Schedule Training Topics	97
Appendix E – Schedule Management Reference Card	98
Appendix F – Schedule Management Plan Template	99
Appendix G – Schedule Assessment Checklist	100

Figures and Illustrations

Figure	Title	Page
1-1	Program/Project Life Cycle Relationships	2
2-1	Vertical Traceability of Schedule Data	7
2-2	Receivables/Deliverables Schedule Illustration	12
4-1	Life Cycle Costs and Phases of Development	19
4-2	Work Breakdown Structure (WBS) Example	20
4-3	Organizational Breakdown Structure (OBS) Example	22
4-4	Responsibility Assignment Matrix (RAM) Illustration	22
4-5	Relationship Between Project Funding and Project Budget	23
5-1	Project Phase / Schedule Detail Relationship	28
5-2	Examples of EVM Measurement Methods	31
5-3	Schedule Coding Crosswalk Example	33
5-4	Task and Resource Calendar Settings	39
5-5	Resource Pool Example	42
5-6	Resource Pool Associated Data Elements	43
5-7	Schedule Resource Loading	44
5-8	Summary Level Cost/Schedule Correlation Check Example	49
7-1	Schedule Insight Penetration Mapped on Risk Cube	56
7-2	Schedule Health Check Example	58
7-3	Critical Path Example	60
7-4	Schedule Performance and Work-Off Trend Example	61
7-5	Cum Baseline vs. Actual Task Finishes & Baseline Execution Rate (BER)	62
7-6	Schedule Performance Efficiency Analysis Example	62
7-7	Linear Projection of "Actuals" Based on Schedule Performance	63
7-8	Total Slack Trend Based on Sched. Performance Example	63
7-9	Baseline Schedule vs. Current Schedule Example	64
7-10	Schedule Variance Report Example	65
7-11	Schedule Margin Log Example	66
7-12	Resource Loaded IMS with Resource Conflicts Example	67
7-13	Resource Loaded IMS with Leveling to Resolve Conflicts	68
7-14	Summary Level Cost/Schedule Correlation Check Example	69
7-15	Schedule Risk Assessment Example	70
7-16	Schedule Performance Insight Using EVM Metrics	72
7-17	Schedule Analysis Using EVM Indicators	73
7-18	Schedule Analysis Utilizing Total Float / SPI	74
8-1	Example Baseline Change Request for IMS	77
9-1	Schedule Performance Reporting	80

Preface

P.1 Purpose

The purpose of schedule management is to provide the framework for time-phasing, resource planning, coordination, and communicating the necessary tasks within a work effort. The intent is to improve schedule management by providing recommended concepts, processes, and techniques used within the Agency and private industry.

The intended function of this handbook is two-fold: first, to provide guidance for meeting the scheduling requirements contained in NPR 7120.5, *NASA Space Flight Program and Project Management Requirements*, NPR 7120.7, *NASA Information Technology and Institutional Infrastructure Program and Project Requirements*, NPR 7120.8, *NASA Research and Technology Program and Project Management Requirements*, and *NPD 1000.5, Policy for NASA Acquisition*. The second function is to describe the schedule management approach and the recommended best practices for carrying out this project control function. With regards to the above project management requirements documents, it should be noted that those space flight projects previously established and approved under the guidance of prior versions of NPR 7120.5 will continue to comply with those requirements until project completion has been achieved.

This handbook will be updated as needed, to enhance efficient and effective schedule management across the Agency. It is acknowledged that most, if not all, external organizations participating in NASA programs/projects will have their own internal schedule management documents. Issues that arise from conflicting schedule guidance will be resolved on a case by case basis as contracts and partnering relationships are established. *It is also acknowledged and understood that all projects are not the same and may require different levels of schedule visibility, scrutiny and control. Project type, value, and complexity are factors that typically dictate which schedule management practices should be employed.*

P.2 Applicability

This handbook provides schedule management guidance for NASA Headquarters, NASA Centers, the Jet Propulsion Laboratory, inter-government partners, academic institutions, international partners, and contractors to the extent specified in the contract or agreement.

P.3 Authority

NPD 1000.0A, NASA Governance and Strategic Management Handbook
NPD 7120.4C, Program/Project Management
NPD 1000.5, Policy for NASA Acquisition
NPR 7120.5, NASA Space Flight Program and Project Management Requirements
NPR 7120.7, NASA Information Technology and Institutional Infrastructure Program and Project Requirements
NPR 7120.8, NASA Research and Technology Program and Project Management Requirements

P.4 References

NPD 1000.0A, NASA Governance and Strategic Management Handbook
NPD 7120.4C, Program/Project Management
NPR 7120.5, NASA Space Flight Program and Project Management Requirements
NPR 7120.7, NASA Information Technology and Institutional Infrastructure Program and Project Requirements
NPR 7120.8, NASA Research and Technology Program and Project Management Requirements
NASA/SP-2010-3404, NASA Work Breakdown Structure (WBS) Handbook
ANSI/EIA-748, Earned Value Management Systems
Academy of Program/Project & Engineering Leadership (APPEL) Website (http://appel.nasa.gov)

Acknowledgments

Primary point of contact: Kenneth W. Poole, Office of Strategic Analysis and Communication, Marshall Space Flight Center.

The following individuals were active participants in the NASA Earned Value Management Scheduling sub-team and are recognized as core contributors to the content of this handbook:

Lynne Faith, (formerly) Dryden Flight Research Center
James H. Henderson, (formerly) NASA/Kennedy Space Center
Kristen C. Kehrer, NASA/Kennedy Space Center
Almond H. Kile, NASA/Ames Research Center
Walter Majerowicz, ASRC, ASRC Aerospace/Goddard Space Flight Center
Kenneth W. Poole, NASA/Marshall Space Flight Center
James G. Smith, Smith & Associates, LLC/Marshall Space Flight Center
Anita M. Thomas, NASA/NASA Headquarters
Lynn L. Wyatt, Vantage Systems, Inc./Goddard Space Flight Center

A special acknowledgement to the following individuals who were primary contributors to the "Scheduling Best Practices Guide" previously created at the Marshall Space Flight Center. This document served as the foundation for much of the content contained in this handbook.

Jimmy W. Black, NASA/Marshall Space Flight Center
Anthony R. Beaver, NASA/Marshall Space Flight Center
Cheryl A. Kromis, MEI – Boeing/Marshall Space Flight Center
John A McCarty, SAIC/Marshall Space Flight Center
Michael W. Norris, Jacobs Engineering Group/Marshall Space Flight Center
Steven O. Patterson, NASA/Marshall Space Flight Center
Kenneth W. Poole, NASA/Marshall Space Flight Center
Donnie E. Smith, MTS/Marshall Space Flight Center
James G. Smith, Smith & Associates LLC/Marshall Space Flight Center
Jeanette C. Tokaz, Jacobs Engineering Group/Marshall Space Flight Center

Chapter 1: Introduction

1.1 Background

This chapter provides an introduction to key elements of NASA's strategic framework for managing programs and projects. Subsequent chapters deal with best practices in how to most effectively administer and satisfy the scheduling requirements that are established in NPR 7120.5, *NASA Space Flight Program and Project Management Requirements,* NPR 7120.7, *NASA Information Technology and Institutional Infrastructure Program and Project Requirements,* and NPR 7120.8, *NASA Research and Technology Program and Project Management Requirements.*

1.1.1 NASA's Program/Project Life Cycle Management Process

NASA programs and projects vary significantly in scope, complexity, cost, and criticality; however, all have a life cycle that is divided into the following four-part management process:

- Formulation – The identification of how the program or project supports the Agency's strategic needs, goals, and objectives; the assessment of feasibility, technology and concepts; risk assessment, team building, development of operations concepts and acquisition strategies; establishment of high-level requirements and success criteria; the preparation of plans, budgets, and schedules essential to the success of a program or project; and the establishment of control systems to ensure performance to those plans and alignment with current Agency strategies.

- Approval (for Implementation) - The acknowledgment by the Decision Authority that the program/project has met stakeholder expectations and formulation requirements, and is ready to proceed to implementation. By approving a program/project, the Decision Authority commits the budget resources necessary to continue into implementation. Approval (for Implementation) must be documented.

- Implementation - The execution of approved plans for the development and operation of the program/project, and the use of control systems to ensure performance to approved plans and continued alignment with the Agency's strategic needs, goals, and objectives.

- Evaluation - The continual, independent (i.e., outside the advocacy chain of the program/project) evaluation of the performance of a program or project and incorporation of the evaluation findings to ensure adequacy of planning and execution according to plan.

For most NASA projects the life cycle management parts for formulation and implementation are further divided into incremental phasing (see Figure 1-1) that allows managers to assess management and engineering progress. The Program/Project life cycle management process receives oversight from NASA Headquarters and a Governing Program Management Council (GPMC).

During the life cycle management process, the following documents set the schedule requirements for NASA programs/projects:

- ***Program/Project Formulation Authorization Document (FAD)*** – The FAD is authorized by NASA Headquarters as the formal initiation of formulation. It identifies the resources, scope of work, period of performance, goals, and objectives for the formulation sub-process.

- ***Program/Project Commitment Agreement (PCA)*** - The PCA is the agreement between NASA Headquarters and the program/project managers that documents the Agency's commitment to implement the program/project requirements within established constraints. It identifies key program/project milestones for the implementation sub-process.

- ***Program/Project Plan*** – This plan is an agreement between NASA Headquarters, the Center Director, and the Program/Project Managers that further defines the PCA requirements and establishes the plan for program/project implementation. It identifies additional key program/project milestones and lower level schedules, and establishes the program/project strategy for schedule development, maintenance, and control.

- ***Program/Project Schedule Management Plan (SMP)*** – While not required by NPR 7120.5 or NPR's 7120.7 and 7120.8, this document is a recommended best practice and may be required by the Program/Project Manager. Refer to Appendix F "Schedule Management Plan Template" for suggested format and content. *(Note: The SMP may be established as a standalone document or as a specified section within the Program/Project Plan. The frequency of this document should be kept consistent with update requirements established for Program/Project Plans in NPR 7120.5. Selected portions of information contained in a standalone SMP will also be required content in a Program/Project Plan.)*

Figure 1-1 depicts the relationship between the program/project life cycle management process and the above documents that establish schedule baseline information.

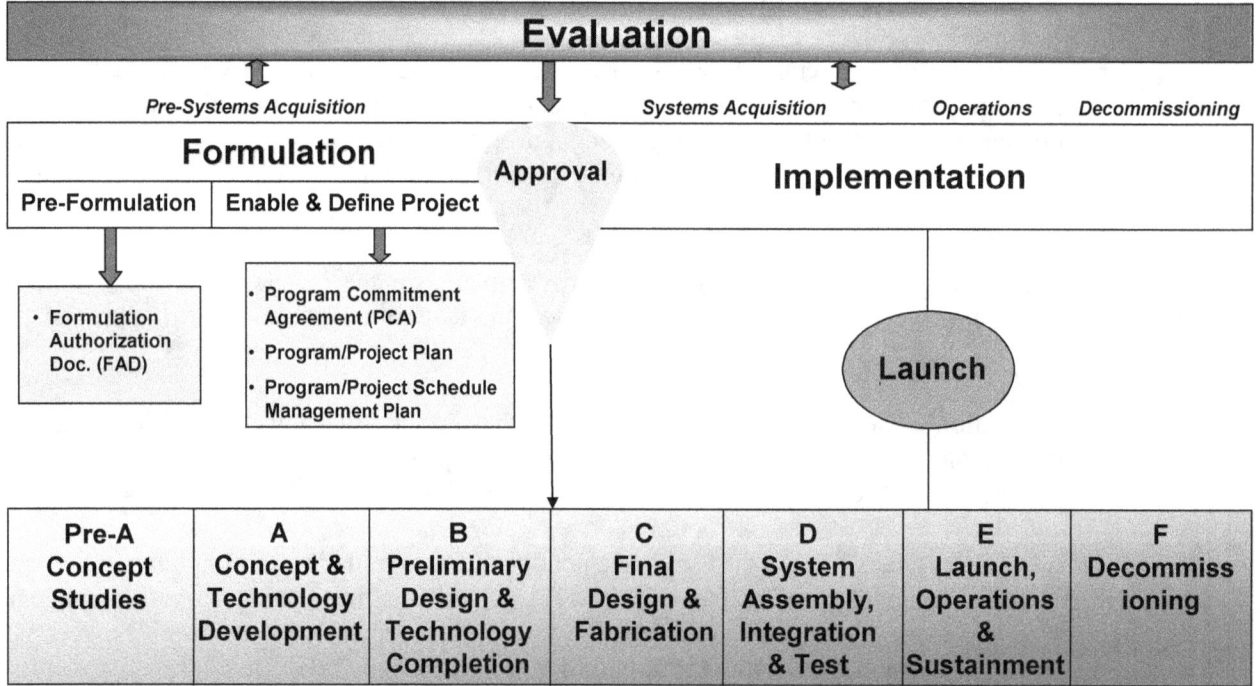

Figure 1-1: Program/Project Life Cycle Relationships

A common thread running throughout the four-part management process shown above is the critical requirement to develop and maintain a Master Schedule, at some level of detail, that clearly defines the necessary effort to be implemented in order to achieve overall mission success. With this management backdrop established, this handbook will provide the necessary guidelines and recommended practices to be used for ensuring schedule management is adequately and consistently implemented on each project across the Agency. Additionally, it is the purpose of this document to draw attention to the critical need for establishing project schedules that are fully integrated with the planned budget and technical content of each project as is recommended by industry "best practices" for effective integrated project management and control.

1.2   Policy Implementation

Sound schedule management requires the establishment, utilization, and control of a baseline master schedule and its derivative schedules. An Integrated Master Schedule (IMS) provides the framework for time phasing and coordinating all project efforts into a master plan to ensure that objectives are accomplished within project or program commitments. Requirements that dictate IMS development, use, and control are found in NPR 7120.5, NPR 7120.7, and NPR 7120.8. It is the responsibility of each project manager and their project team to ensure that these schedule management requirements are adhered to, not only during initial IMS development, but also in the on-going updating and maintenance.

All schedule management requirements contained in the policy documents listed above are generally consistent with industry standards. The remainder of this handbook defines the recommended best practices for fulfilling the schedule management requirements set forth in NPR 7120.5, NPR 7120.7, and NPR 7120.8.

Chapter 2: Schedule Management Overview

2.1 Schedule Management Approach

Schedule management encompasses the development, maintenance, control, and archival of the IMS. The IMS constitutes the basis for time phasing and coordinating all program/project effort to ensure that objectives are accomplished within approved commitments. Integrated schedules are crucial for all levels of management oversight within NASA and its contractor community. Program level master schedules should contain the necessary tasks and milestones that reflect the total integrated plans for each project within the program. While it is true that all program work scope must be included within a program IMS, the levels of detail may vary to accommodate the specific management needs established at the program level. Schedule management at the project level entails the creation of an IMS that contains integrated tasks and milestones for both contractor effort and all effort being worked by in-house NASA government personnel. Regardless of the type project being implemented it is critical that the IMS contain data addressing the total scope of work at a consistent level of detail to allow for discrete progress measurement, management visibility, and critical path identification and control. This approach will allow Program/Project Managers greater visibility and capability to adequately plan the necessary resources, and to ensure adequate budget will be available to accomplish the work when it is planned.

The following are the schedule management process groups contained in this handbook: Pre-Schedule Development, Program and Project IMS Development, Status Updates and Schedule Maintenance, Schedule Assessment and Analysis, Schedule Control, Schedule Reporting, and Schedule Data and Lessons Learned Archival. These process groups are described in more detail in subsequent chapters. These process groups interact with each other and with other project management processes such as cost estimating, change control, and risk management.

2.1.1 Program Integrated Master Schedule

The Program Integrated Master Schedule (PIMS) is a core tool for the integration, control, and analysis of all program work scope. The PIMS contains tasks, milestones, and interdependencies configured in a manner that accurately models the implementation plan for all approved scope contained in the program. Program scope includes both the authorized work within each project contained in the program, and also all authorized effort that exists only at the program level.

Prior to developing a PIMS, careful consideration should be given to several important factors that will impact its purpose and capabilities for future use by the program team. Key factors to be considered include, but are not limited to the following:

- The levels of schedule insight and analysis capability that are desired throughout program implementation
- The magnitude and complexity of schedule data to be maintained and processed
- The schedule management tools that potentially will be used
- Potential integration requirements between the IMS, the NASA Core Financial System, and a designated Earned Value Management (EVM) reporting tool.

Program management should understand that the level of program insight and analysis that can be achieved is heavily dependent on the level of detail contained in the PIMS. It should also be understood that detailed critical path identification and analysis, as well as detailed insight into program issues cannot be done with only summary-level schedule content. Therefore, the level of schedule detail contained in the PIMS is very important to accomplishing the established program insight and analysis processes. It is also very important that the amount of schedule data, along with the complexity of processes required to maintain and analyze the data must be balanced against the number and skill levels of the personnel responsible for it. It is clear that effective program oversight cannot be done without adequate numbers of staff to accomplish it. Choosing the right schedule management tools is also very critical to success, not only for the program team, but also for the project teams, and the contractors that potentially are involved. It should be clearly understood during up-front program planning and process development that there are numerous management tool sets available that do not allow for easy and/or accurate transfer and integration of schedule and performance data. It is crucial for achieving successful program management that tool sets which provide efficient and accurate transfer and integration of data be chosen, and where possible, mandated for all program participants.

With the above considerations in mind, the following options are offered as recommendations for PIMS development:

- The preferred strategy recommended for program use in developing a PIMS is to create a schedule dataset that, at a minimum, integrates the full detailed IMS schedules for each project included in the program. If inter-relationships exist between any of these projects, then appropriate logic relationships should be included to accurately model those interdependencies. Additionally, all tasks and milestones reflecting effort to be implemented only at the program level must also be included, as well as all program-level control milestones that have been established. The approved program/project WBS element coding should also be included in the PIMS. This strategy provides the overall capability for integrated insight and oversight of all program work, including detailed critical path and program issue information. It should be understood, however, that while this strategy is the preferred technique, it may not be a practical approach for the PIMS. As stated earlier, several associated factors as explained above, may prohibit this approach from being used.

- An alternate technique that can be used for PIMS development, when the preferred strategy is not practical, is to integrate summary versions of each project IMS schedule. When using this technique it is important to keep the level of summarization consistent with the desired level of insight, as influenced by risk, cost, and criticality (both schedule and technical). All projects, as with all WBS elements, won't necessarily require the same level of program insight. Interdependency relationships for all summary tasks and milestones should be established and maintained. Inter-project logic relationships, as well as WBS element coding should again be identified. When using this alternate technique, it is still recommended that all tasks and milestones reflecting effort being implemented only at the program level be included in detailed fashion to allow for adequate program team oversight and management control of the work they are responsible for. Caution should be exercised in using this approach when performing critical path identification and analysis. Summary-level schedule data will not typically identify many of the detailed integration points needed for accurate task sequencing. The resulting impact is that accurate critical path logic flows cannot be identified. This potentially leads to erroneous summary-level critical path information which is not accurate or consistent with the detailed critical path information. It must be acknowledged and understood by the program management team that when using summary-level project IMS data within the PIMS, the level of insight, control, and analysis will also have to be raised to a higher level. This means that the program team will have to depend more heavily on the detailed schedule insight and analysis provided by the project teams involved in the program.

- A third approach that can be employed for PIMS development is using milestone sets to reflect the major events in accomplishing all program effort. Sets of meaningful event milestones reflecting each project's scheduled effort would be used in-place of detailed or summary-level tasks. Milestone interdependencies are much more difficult to reflect accurately when using this technique. This difficulty is due to the method in which the planner/scheduler (P/S) has to account for the effort being carried out in between the milestones. In order for the PIMS to keep the proper time-phasing for the numerous project milestones the P/S must either incorporate appropriate schedule lag values between each milestone, or assign date constraints to each milestone included in the schedule. Both of these practices are not conducive to sound schedule analysis. If maintaining logic interdependencies for critical path analysis, or having insight into sequence issues are important to the program team then this PIMS strategy is not recommended. However, if developing a program-level schedule that only requires maintaining a picture of what and when major project events are to happen, then this approach is a viable solution. Even when utilizing this approach, it is still recommended that all tasks and milestones reflecting effort being implemented only at the program level be included in detailed fashion. This allows for adequate program team oversight and management control of the work they are responsible for.

When considering program-level schedules, it should also be noted that within NASA, programs are categorized into the following four groupings:

- **Single-project programs**, such as the James Webb Telescope program, have only one project that makes up the program. For this type, the PIMS will most likely not have interdependencies to other projects.
- **Tightly coupled projects programs**, such as Constellation, contain projects that have a high degree of organizational, programmatic, and technical commonality. This type of program requires a much higher degree of integration between the projects potentially resulting in numerous inter-project interdependencies in the PIMS.
- **Loosely coupled programs**, such as Mars Exploration, contain projects that have organizational commonality, but little programmatic or technical commonality. These projects will typically have minimal or no inter-project interdependencies in the PIMS.
- **Uncoupled programs**, such as Discovery, contain projects that are implemented under a broad scientific theme and/or a common implementation concept, but each project will be independent of other projects in the program. These projects will have minimal or no inter-project interdependencies in the PIMS.

2.1.2 Project Integrated Master Schedule

The project-level IMS, as defined in Appendix B, is "an integrated schedule developed by logically networking all detailed project activities. The highest level schedule is the Master Schedule supported by Intermediate Level Schedules and by lowest level detail schedules." The IMS serves as the basis for planning and performance insight for all project effort. The IMS provides the management vehicle which enables integration of the approved project work scope reflected in the work breakdown structure (WBS), budget, and certain project risks. The IMS typically reflects both baseline and current schedule data. This is to say, the IMS will reflect both the project approved time-phased plan *(including all subsequent approved changes),* and the time-phased plan with its current task progress, sequence, and forecasts. The time-phasing of tasks provided by the IMS is critical to successful implementation of

Earned Value Management (EVM) and the development of a Performance Measurement Baseline (PMB). See Chapter 5 for the schedule development processes and best practices.

The IMS provides the program/project manager a single integrated source of schedule data that accurately reflects how the planned work is to be implemented. This dataset will be maintained in an automated schedule management tool and consist of tasks/milestones, task durations, interdependencies, project constraints, contractor data, and an assigned data coding structure. Using the assigned coding structure, the scheduling tool is able to filter and summarize schedule data to provide reports at the summary Master Schedule, intermediate, and detail schedule levels. It is important to note that all levels of schedule reporting should be provided from a single IMS dataset and not from separate schedule sources. Vertical traceability from detailed tasks to higher level program/project milestones should exist as indicated in Figure 2-1. Detailed schedules, as reflected in the figure below, contain the lowest level planning segments. These segments generally correlate to groupings of associated work and are typically reflected as detailed tasks and milestones which may be summarized to a level of detail called work packages. Groupings of work packages may also be summarized to a higher detail called control accounts. While different techniques and formats may be employed within detailed schedules, it is critical that tasks be defined at a low enough level to allow for finish-to-start interdependency relationships where feasible, accurate progress measurement, issue identification, and correlation to higher level milestones.

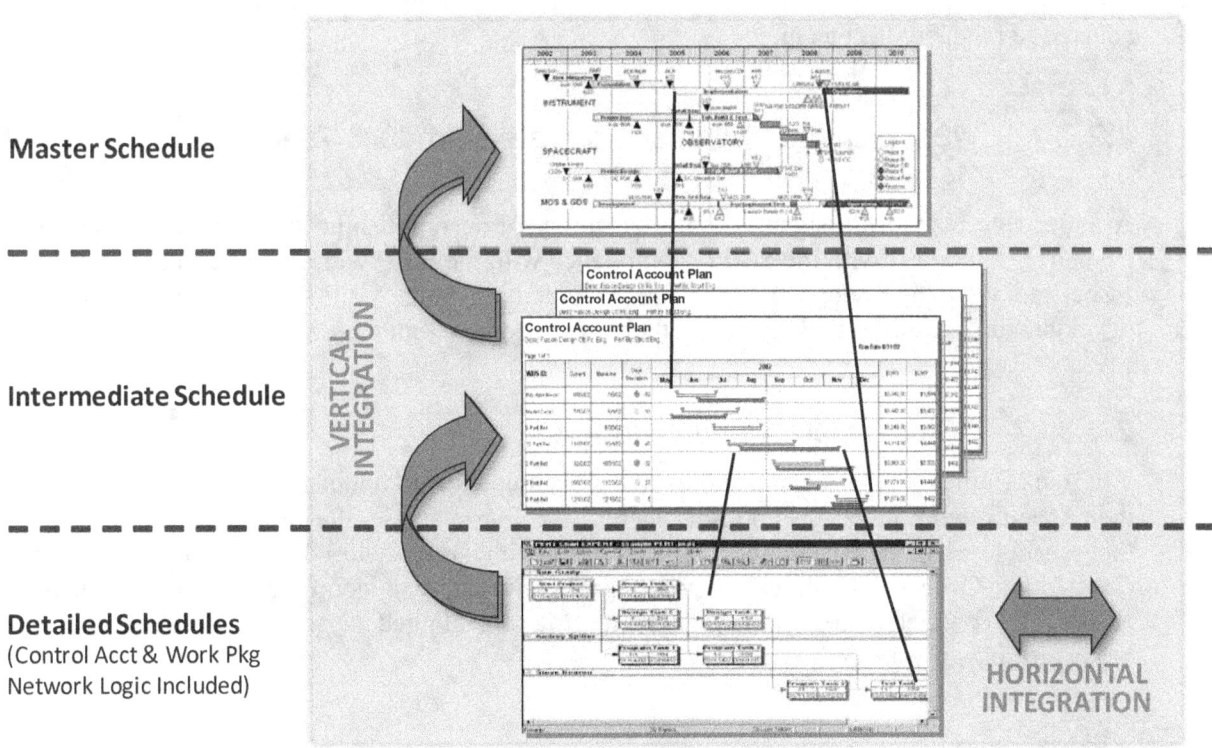

Figure 2-1: Vertical Traceability of Schedule Data

2.1.3 Schedule Management Plan

A Project Schedule Management Plan (SMP) should be prepared during Phase A of Project Formulation prior to Key Decision Point (KDP) B for each project. The SMP will describe and define the techniques and methods to be used in implementing schedule management processes. The SMP can be a stand-alone plan or a subsidiary component of the Project Plan. Regardless of how it is structured within a project's documentation, the Project SMP should be subject to document control. The SMP is not intended as a detailed procedure for performing scheduling. Rather, it is a guideline for applying generally accepted project scheduling practices. See Appendix F for a Project SMP Outline Template.

The content of the IMS and the overall SMP approach should be dependent upon how a project is organized. For example, there could be in-house, prime contractor, and/or external partnership activities which will influence the planning process. Additionally, schedule management should be in accordance and integrated with the institutional EVM processes and methodologies on programs/projects.

2.2 Roles and Responsibilities

2.2.1 Project Managers

The Project Manager's (PM) role is to ensure schedule management processes are applied in a manner that supports the project's requirements and objectives. The PM is responsible, with support of his team's project control personnel, for the schedule development guidelines, IMS development, plan approval, schedule execution, and baseline control of the IMS, and ensuring the institutional processes and procedures, necessary resources, and tools and techniques are applied to ensure requirements are adhered to by the project team.

2.2.2 WBS Element Owners/Integrated Product Team Leads/Control Account Managers

A major role for those assigned the responsibility of accomplishing the work contained in each WBS element is to comply with the SMP. Compliance with the SMP will help to ensure that the deliverables associated with their scope of work are provided on time. WBS Owners are responsible for the development, execution, and control of their work scope within the IMS.

2.2.3 Other Project Team Members

The role of other project team members and stakeholders is to understand the IMS and how it relates to their specific work processes and responsibilities. For example, the finance team will coordinate with the schedule team to ensure the budget phasing integrates with the IMS timeline. Similarly, the Contracting Officer's Technical Representative (COTR) will coordinate with the schedule team to ensure the contractual deliverables and the IMS are aligned (e.g., data deliverables, reviews, and hardware and software deliveries).

2.2.4 Planner/Scheduler

The Planner/Scheduler (P/S) role is to implement SMP processes in order to ensure the project's objectives are successfully achieved. The P/S must be familiar with the project technical scope and able to translate that information into the IMS network logic model. The P/S accomplishes this, in part, by providing: 1) planning through coordination with the project team to define project requirements and schedule objectives, and to develop the IMS; 2) analysis and insight to the project team by reporting schedule progress, performance, variances, and forecasts, evaluating risks, and performing "what-if" analysis; and 3) IMS control by assisting the project team to manage change to the IMS which includes baseline change control. The P/S is also responsible for utilizing project management software tools and techniques to develop, maintain, and control the IMS. Finally, project P/S(s) must be able to communicate and coordinate effectively with all members of the project team, be proactive in their approach to problem solving, and to understand project management processes (Initiating, Planning, Executing, Reporting, Controlling, and Closing).

2.3 Contractor Schedules and Coordination

2.3.1 Contractor Schedule Management and Coordination

The P/S should coordinate with the responsible COTR to develop the schedule management and reporting requirements for applicable procurements. These requirements may be contained in the contract Statement of Work (SOW), Contract Data Requirements List (CDRL), and/or Data Requirements Document (DRD). The objective is to obtain the schedule information necessary to manage the IMS and enable informed decision making. SOW, CDRL, and DRD should be structured in order to take maximum advantage of contractors' existing scheduling systems, capabilities, and formats. Additionally, this Handbook may be provided as guidance to contractors and subcontractors.

The SOW, CDRL, and DRD must provide clear requirements for contractors in the areas of scope content, deliverable expectations, and data requirements in order to avoid confusion during project implementation. To effectively integrate contractor schedule data into the project IMS it is imperative that a clear understanding exists between the government and contractors about such details as schedule content, level of detail, formats, reporting frequency, tools, thresholds, responsibilities, and controls. Anything that can be done during project initiation to clarify what is expected of the contractor will reap huge benefits in saving time and money, and reduce stress and frustration levels in the personnel carrying out project implementation. This approach will also serve to provide additional risk mitigation throughout the project life cycle.

The contract SOW for schedule management should clearly delineate the work and deliverables that are to be scheduled, the type of schedule products to be provided, the DRD to be followed, and any special considerations required for carrying out the contracted work. The CDRL is a listing of the technical information and reports required for a contract including submittal and approval criteria and instruction. The IMS DRD is a document that provides specific requirements for schedule content, level of detail, format, reporting frequency, applicable thresholds, and guidance for variance rationale. Appendix C to this document contains an example DRD for schedule deliverables.

2.3.2 Monitor, Assess, and Control Performance of Contractor Schedules

Contractor schedules must be continually monitored, assessed, and managed to assure successful project performance. To effectively accomplish each of these functions requires the contractor, on a monthly basis, to electronically submit their project IMS database electronically to NASA's project office in its native or equivalent file format (e.g., MS Project, Primavera, Open Plan, MPX, XML, and other). Having access to the IMS database in its native file format makes it possible for the government P/S to monitor, assess, and evaluate, at any level of detail, the quality and integrity of its task sequencing, projected dates, primary and secondary critical paths, assigned constraints, resources, coding, structure, and current status. Results of this type of detailed, on-going evaluation should be made available to the contractor for consideration and/or correction. This approach enables the government team to partner with the contractor in identifying potential schedule risks and selecting the best strategies for mitigation.

2.3.3 Maintain Baseline Review and Integrity of Contractor Schedules

Helping the contractor to establish and maintain credible schedule baselines is integral to sound project schedule management and performance measurement. Schedule baselines must be approached as a joint government/contractor project team product. While the contractor's P/S typically produces and maintains most of the schedule products provided, it is the whole project team, including both government and contractor members, that must claim ownership in the schedule content and its validity. A schedule management process should be implemented within each contractor's effort which dictates that prior to schedule baselining, the responsible COTR (and/or his designated representatives) and all contractor integrated product/project team (IPT) or WBS element engineers and managers must perform a thorough IMS review. This review should cover not only schedule content, but also task/milestone sequencing, resources, slack (float) analysis, probabilistic schedule uncertainty, and all valid constraints that apply. This type of review should be carried out prior to initial project baselining and then repeated after major project changes or at specified intervals as-directed by the NASA customer and accompanied by project management buy-in to ensure on-going schedule integrity.

For contracts with Earned Value Management (EVM) requirements an Integrated Baseline Review (IBR) must be conducted. The initial IBR should be accomplished within the first three to six months after contract award or after approval of the Project Plan by the Mission Directorate Associate Administer (MDAA), or Mission Support Office Director (MSOD). On projects of short duration, it may be necessary to initiate the review earlier in order to make best use of the information derived from the review. During this review joint contractor and government teams review the total project cost, schedule, and technical baseline for the purpose of ensuring that a valid baseline is in place and that a mutual understanding and agreement exists in the scope of work and in the amount of resources required. An IBR additionally identifies and alerts the project team to potential project risks.

2.4 In-House Schedules and Coordination

The P/S(s) should support the Technical Managers/Leads for all scheduling requirements and perform planning coordination and detailed scheduling for all in-house projects in partnership with other organizations as required. In-house effort could include the entire project, or a major element within the project such as the spacecraft, a single scientific instrument, or systems integration and test. *The role of support service contractors would also be considered part of in-house effort.*

2.4.1 Establish the Schedule Management Plan for In-House Projects

During Phase A of an in-house project, the project P/S is responsible for developing a Schedule Management Plan (SMP) for the project. The SMP should contain the schedule development, maintenance, and control processes that should be implemented on the IMS. This plan will also describe the schedule management tools that will be utilized within the specific projects. The SMP should also be reviewed, approved, and signed-off by the appropriate Project Manager.

2.4.2 Develop the Integrated Master Schedule for In-House Projects

For in-house projects, the P/S(s) should coordinate with the responsible Technical Leads to develop the IMS. There are four approaches to compiling, integrating, and formatting the provided schedule data when developing the IMS. The first, and most desired, approach is *Fully Automated Schedule Integration* by either incorporating the in-house organizational or WBS element schedule as a sub-network into a single schedule file or using inter-project linking capabilities in the scheduling software tool. Either integration technique provides an automated means to logically connect the schedule dependencies among all project work elements. This approach requires careful planning early in the formulation stage and coordination and follow-up with full integration throughout the implementation stage. This approach provides the project management team with the maximum insight capability into the IMS plan, performance and forecast.

A second approach that may be required when developing the in-house IMS is the *Partially Automated Schedule Integration* method. This approach offers some flexibility to deviate from the fully automated integration methodology for IMS development and is used when the automated schedule data provided is not appropriately formatted or defined. For example, some in-house developed schedules may not be compatible for use "as is" in the IMS envisioned with *Fully Automated Schedule Integration*. In these cases, the P/S(s) should coordinate with the Technical Leads to create sub-networks of the in-house schedule(s) within the IMS. These sub-networks, in Critical Path Method (CPM) format, must identify: a) the contributing organization's work scope, b) key deliverables and c) critical schedule dependencies with other organizations. In these cases, the P/S(s) must ensure that the sub-networks created for incorporation into the IMS integrate satisfactorily into the in-house organization's schedule for the baseline, actual performance and current forecast. The *Partially Automated Schedule Integration* approach also provides critical path, "what-if" and other forms of schedule analysis.

A third approach that may be required is the *Master Logic Network* technique and is employed when an in-house organization is unable to provide meaningful schedules to the project office. In these cases, the project P/S(s), in coordination with the contributing organizations and Technical Leads, develop schedule estimates that best reflect the schedule plan, performance and forecast for the contributing organization(s) in question. This may involve preparing the actual schedules for them. In other instances, it may necessitate the creation of in-house schedules based on whatever information is available. This approach provides the project team with a basic IMS for the project, and also supports a limited critical path and "what-if" schedule analysis capability.

If any of the first three approaches to an IMS at the project-level are not viable, or the contributing in-house organization(s) are unable to provide useful project schedules to the project scheduling office, the *Receivables/Deliverables Matrix* technique can be used. With this approach, no attempt is made to integrate all of the project schedules in an automated or electronic fashion. Instead, schedules from contributing organization(s) are individually tracked in the contributing organization's "native" format or other format that best portrays the schedule plan, performance and forecast. The *Receivables/Deliverables Matrix* is the mechanism to achieve a rudimentary form of schedule integration in these instances. This matrix is a simplified spreadsheet that lists all key schedule deliverables or interface points among the organizations supporting the project. The matrix also includes the baseline and current projection for the need date of the deliverable by the project, as well as the baseline and current projection for the delivery date of the deliverable provided by the in-house or contractor organization(s). This approach provides minimal insight into the critical path, schedule performance measurement and schedule analysis. Nevertheless, if rigorous configuration control of the *Receivables/Deliverables Matrix* is maintained, and the data is accurately integrated with the schedule data (see Figure 2-2) that the contributing organization(s) provide, some basic level of an IMS is achieved at the project level. Use of this approach should require the concurrence of the Project Manager and a justification as to why none of the other approaches to the IMS are feasible.

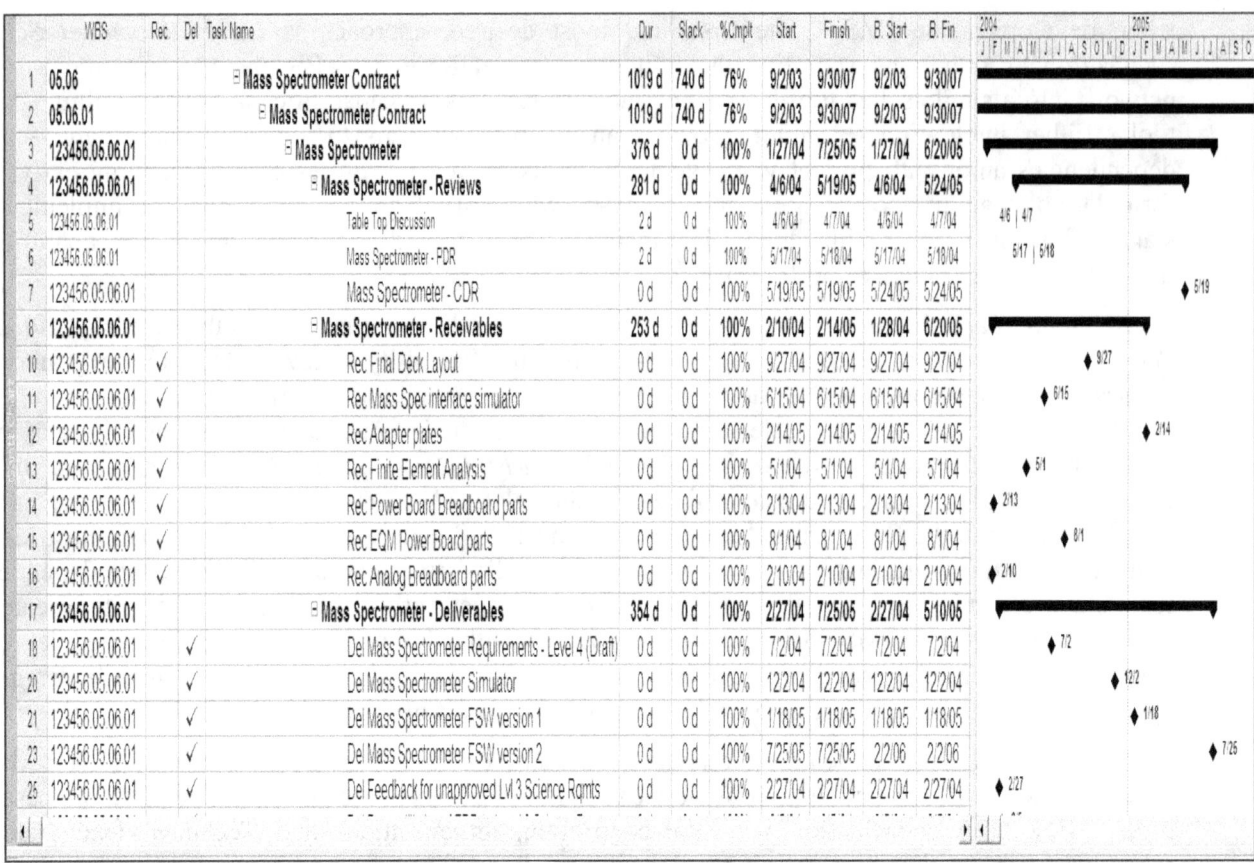

Figure 2-2: Receivables / Deliverables Schedule Illustration

2.4.3 Monitor, Assess, and Control Performance of In-House Schedules

In managing NASA project schedules, the in-house NASA-developed schedules must be continually monitored, assessed, and managed in a manner similar to that identified for contractor schedules. The IMS, even though developed by NASA personnel, must follow the same guidelines and best practices as would be expected from a contractor. A sound logic network-based schedule must provide the basis for all schedule data provided to management for critical project decisions. Utilizing the IMS database in its native file format makes it possible for the assigned project P/S to effectively monitor, assess, and evaluate, at any level of detail, the quality and integrity of its task sequencing, projected dates, primary and secondary critical paths, assigned constraints, resources, coding, structure, and current status. Results of this type of detailed, on-going evaluation and analysis should be made available to the appropriate project managers and responsible team members for consideration and/or correction. This approach enables the total integrated project team to more effectively identify potential schedule risks in a timely manner and to select the best strategies for mitigation.

2.4.4 Maintain Baseline Review and Integrity of In-House Schedules

Establishing and maintaining credible in-house schedule baselines is integral to sound project schedule management and performance measurement. Schedule baselines must be approached as an integrated project team product. While the in-house project P/S typically produces and maintains most schedule products, it is the total project team that must claim ownership in the schedule content and its validity.

A schedule management process should be implemented within each project organization that dictates that prior to schedule baselining, the project manager along with each of the responsible WBS element owners and engineers must perform a thorough IMS review. This review should cover not only schedule content, but also the task/milestone sequencing, associated resources, and all valid constraints that apply. This type of review should be carried out prior to initial project baselining and then repeated as necessary after major project changes or at specified intervals and/or events (i.e., Program Planning Budget Execution (PPBE) updates, Preliminary Design Reviews (PDR), Critical Design Reviews (CDR), etc.) accompanied by project management buy-in to ensure on-going schedule integrity. It must be stressed however, that establishing new IMS baseline schedule dates should only occur when major scope and/or budgetary changes have been encountered and formally approved by program/project management.

2.5 External Organization Schedules and Coordination

2.5.1 Integrate Partner Schedules into Integrated Master Schedule

Schedule reporting requirements provided for partnerships between a project and other NASA centers, research institutions, international partners, or other business arrangements not involving contracts or procurements should be incorporated into a Memorandum of Understanding (MOU), Space Act Agreement, Letter of Agreement (LOA), SOW, SMP, or other appropriate documents. This will enable the IMS to fulfill its intended function as an effective and efficient integrated project management tool. It should be noted that some arrangements permit NASA schedule expertise to be used in the development of partner schedules to facilitate integration for enhanced management capabilities. For example, some Science Mission Directorate (SMD) projects with deliverables provided by universities often provide direct scheduling support to the institution from their own project scheduling staff to assist in schedule development and status reporting.

2.5.2 Monitor, Assess, and Control Performance of External Partner Schedules

Schedules from external partners (e.g., other NASA Centers, other government agencies, and universities) must be continually monitored, routinely assessed, and managed effectively to enhance performance. To accomplish each of these functions requires the external partner to routinely and electronically submit their project IMS database to NASA's P/S in its native file format (e.g., MS Project, Primavera, Open Plan, and other). Having access to the partner's IMS database in its native file format makes it possible for the NASA P/S to monitor, assess, and evaluate, at any level of detail, the quality and integrity of its schedule content, task sequencing, projected dates, primary and secondary critical paths, assigned constraints, resources, coding structure, and current status. Results of this type of detailed, on-going evaluation and analysis should be made available to the external partner for consideration and/or correction. This approach enables the NASA team to partner with all external organizations in identifying potential schedule risks and selecting the best strategies for mitigation.

2.5.3 Maintain Baseline Review and Integrity of External Partner Schedules

Establishing and maintaining credible external partner baseline schedules is integral to sound schedule management and performance measurement. Schedule baselines must be approached as a joint NASA/external partner team product. While the external partner's P/S typically produces and maintains most of the schedule products provided, it is the total NASA and external partner team that must claim ownership in the schedule content and its validity. A schedule management process should be implemented within each external partner organization which dictates that prior to schedule baselining, the responsible NASA project manager, external partner IPT, and/or WBS element owners and engineers must perform a thorough IMS review. This review should cover not only schedule content, but also the task/milestone sequencing, resources, and all valid constraints that apply. This type of review should be carried out prior to initial project baselining and then repeated as necessary after major project changes or at specified intervals and/or events (i.e., Program Planning Budget Execution (PPBE) updates, Preliminary Design Reviews (PDR), Critical Design Reviews (CDR's), etc.) accompanied by project management buy-in to ensure on-going schedule integrity.

2.5.4 Management/Reporting Agreements for External Partner Schedules

Just as NASA documents specify management and reporting requirements for prime contractors, the same must be established and agreed to with external partners. MOU, LOA, and any other binding agreement with an external partner must be structured in a manner that takes full advantage of their existing scheduling systems, capabilities, and formats. Additionally, these documents must provide clear specifics and/or guidance for the external partner in the areas of scope content, deliverable expectations, and data requirements so that minimal confusion arises during project implementation. To effectively integrate the partner's schedule data into the project IMS it is imperative that a clear understanding exists for such details as schedule content, level of detail, formats, reporting frequency, tools, thresholds, responsibilities, and controls. This approach will also serve to provide additional risk mitigation throughout the project life cycle.

The external partner agreement should contain guidance for schedule management that clearly delineates what is to be scheduled, the type of schedules to be provided, the data requirements, and any special considerations required for carrying out the agreed to scope of work. The agreement should also document the specific project deliverables along with specific information on quantities, WBS relationship, due dates, delivery location, means of delivery, and any other pertinent guidance needed by the external partner. External partner agreements should also consider implications and impacts resulting from regulations contained in the Federal Acquisition Regulations (FAR) and the International Traffic in Arms Regulations (ITAR).

2.6 Schedule Training

The success of schedule management depends primarily on the quality of planning and scheduling skills that are dedicated to programs/projects across the Agency. To ensure strong, consistent scheduling expertise is available, it is imperative that the appropriate training be taken by the project team members that are involved in developing, maintaining, using, or analyzing project schedules. See Appendix D for a list of recommended training topics in the area of project schedule management.

Bringing improvement to scheduling expertise at all NASA facilities should be carried out through a four-pronged approach. First, project personnel should have available to them, through each center's training organization, detailed, partial and multi-day introductory, intermediate, and advanced courses in the area of project scheduling. This training may be provided by civil servants and/or procured private vendors. Second, project personnel should also have available to them, through their respective center's training organization, short schedule training overview modules. The overview classes will generally last a half day or less and should be more convenient for project personnel to fit the desired training into their daily schedules. Third, a curriculum of self-taught training material should be available via the System for Administration, Training, and Educational Resources for NASA (SATERN) training website (See https://satern.nasa.gov/elms/learner/login.jsp). Fourth, it is recognized that for classroom training to be effective it should always be accompanied by appropriate on-the-job training (OJT) and/or hands-on training (HOT).

It should be understood that the NASA schedule management training curriculum will cover an evolving, growing list of topics. Training chosen should address the needs and requirements of project management team responsibilities.

Chapter 3: Schedule Management Tool Considerations

3.1 Overview

Schedule management tools are used to develop, maintain, analyze, and control project schedules. Agency-wide adherence to schedule management tool recommendations can improve data sharing capability, increase interoperability with other agency standard tools, and potentially enable speedy accessibility through common Agency-wide procurement vehicles. Prior to selecting a schedule management tool it is recommended that program/project functionality requirements and desires be defined. This approach will help ensure that tools are selected that best satisfy those needs.

3.2 Best Practices

It is a recommended best practice that automated schedule management tools be used on all NASA programs/projects. It is also recommended that these tools satisfy the functional, interface, and technical capabilities listed in the paragraphs below:

3.2.1 Functional Capabilities

Schedule management tools should perform the following functions:

- Provide for entering and editing of baseline plan, current/forecast plan, and accomplished (actual) schedule data
- Specify relational dependencies between tasks and milestones (including lag and lead values as needed, but kept to a minimum)
- Define project calendars that reflect the business schedule (e.g., workdays, non-workdays, holidays, and work-hours)
- Display and print project schedules in Gantt and network diagram form
- Calculate total slack (float) and free slack for all project tasks and milestones
- Provide user-defined code fields for filtering, grouping, summarizing, and organizing data
- Create, view, and print basic reports such as task, cost, and resource listings
- Provide capability for resource loading and leveling

3.2.2 Interface Capabilities

Tools selected for use in the implementation of recommended schedule management practices should satisfy the following interface capabilities:

- Supports data interface to chosen in-house EVM applications (e.g., Internally developed EVM spreadsheets or commercial EVM applications)
- Supports data interface to EVM data analysis applications
- Supports data interface to schedule risk assessment applications

3.2.3 Technical Capabilities

Tools selected for use in the implementation of recommended schedule management practices should satisfy the following technical capabilities:

- Supports email capability of schedule data in native file formats
- Uses Open Database Connectivity (ODBC) or Dynamic Data Exchange (DDE) standards to read/write to other databases
- Provides capability of saving data files such as MPX, DBF, XML, HTML, and X-12
- Provides online HELP capability
- Provides capability for creating PDF files or graphic files such as: jpeg, bmp, gif, or tif.

Chapter 4: Pre-Schedule Development Activity

4.1 Overview

The first steps in developing a new project schedule include understanding the project work scope, developing a Work Breakdown Structure (WBS) and Organizational Breakdown Structure (OBS), understanding project funding dynamics, and reviewing pertinent agreements and authorization project documents. Project schedule content should be consistent with applicable project documentation and requirements.

Project schedules may be the product of both in-house and contractor efforts. The pre-schedule considerations may require the development and tailoring of a schedule data requirements document (DRD) to be included as part of the request for proposal (RFP) or contract. For work performed in-house, a work authorization process should be in place that identifies schedule requirements in addition to the approved budget. The project team should consider the topics and documents listed in section 4.2 "Best Practices."

4.2 Assignment of Project Planner/Scheduler

Pre-scheduling activities should include identifying the resources required for developing and maintaining the project schedule. This includes assigning and training the personnel who will be responsible for developing and maintaining the schedule. This could be a P/S, the project manager assuming the analyst's responsibilities, or some other team member. For the purpose of discussion in this section, the person assigned these responsibilities will be referred to as the P/S.

It is critical that the P/S be assigned early in the life cycle of every program or project. To ensure the information necessary for schedule management and insight is received early, the P/S should be assigned by the start of a project or at least no later than Phase A of the project life cycle. The P/S should participate in the gathering and/or development of schedule documentation and requirements (e.g., reports, studies, authorizing documents, WBS), and when applicable, ensure that the proper DRD(s) are included in the RFP and contract. The P/S should also work to establish a process for schedule status (see section 6.2.1) with team members, as well as a process for controlling changes to the baseline (see section 8.3). Refer to section 2.2.4 for additional roles of the P/S. The P/S should be trained in all areas, including software tools, schedule maintenance and control, and scheduling techniques. Refer to section 2.6 for additional information on schedule training.

4.3 Program/Project Scope

An understanding of the work content must exist before a valid schedule can be developed. The initial step in gaining this understanding is a thorough review of the project scope definition. The P/S(s) should gather relevant data and documentation (e.g., reports, studies, authorizing documents, WBS) for IMS development. If relevant data or documentation has not been developed, the P/S(s) should participate in the development of these documents. It is important to realize that all project personnel may not have the same interpretation or understanding of the approved SOW. Resolving these differences is necessary for the development of an accurate and useable schedule for project management. The P/S can play a significant role in helping to resolve these differences by asking the right questions (e.g., in what WBS element does specific effort belong? What type of deliverable is required? What type of testing is required?) and by bringing to light the areas of conflict so that responsible managers can come to an agreement on the work scope. For example, the P/S should always help the project team understand the necessary inputs (e.g., responsibility, in-house or procured effort, quantities, facility requirements) to task and schedule definition, as well as the inherent interfaces involved.

It is imperative to identify and plan the total work content as early in the planning process as possible. This will help identify potential conflicts that can lead to schedule changes. The later in the project life cycle a required change is identified, the more costly it will be to the project in terms of time and money (see Figure 4-1). Many cost and schedule project risks can be mitigated with better up-front planning.

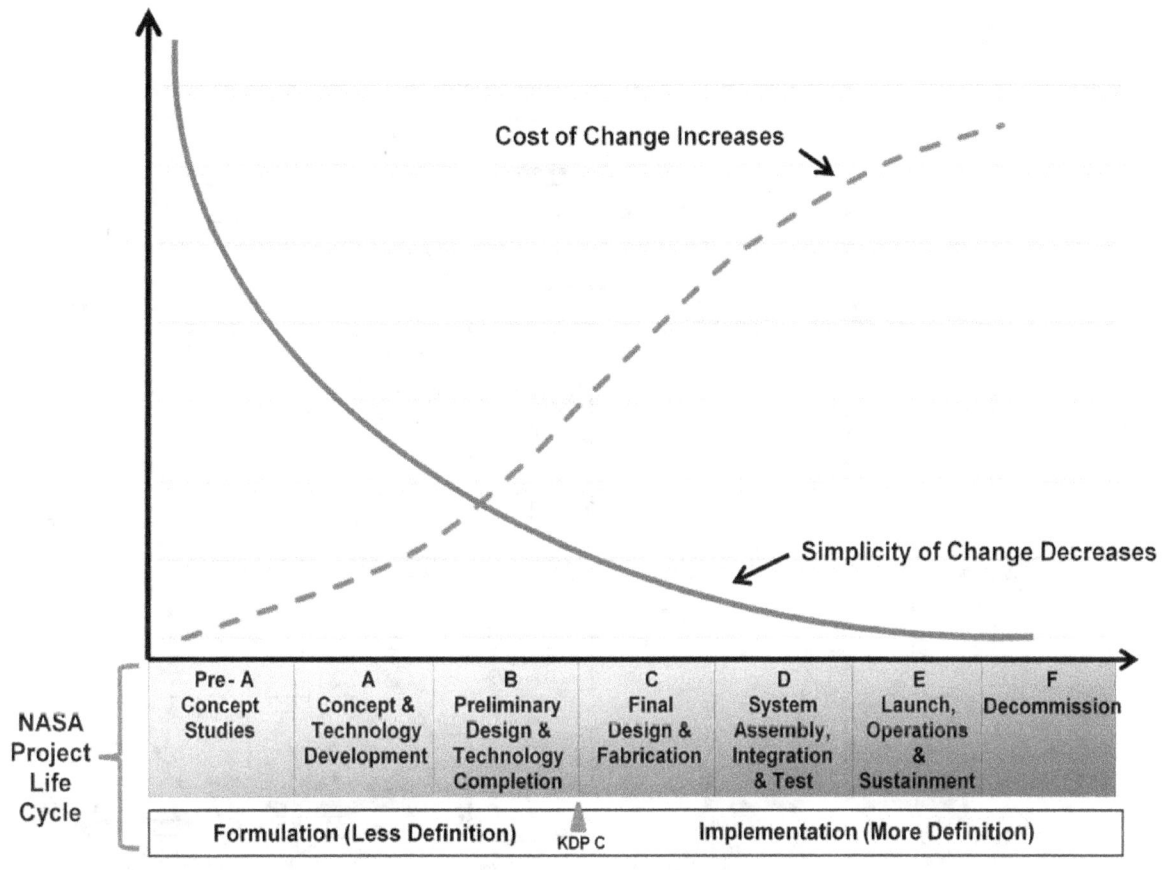

Figure 4-1: Life Cycle Costs and Phases of Development

4.4 Project Work Breakdown Structure (WBS)

Just as an accurate understanding of scope is crucial to the development of a valid and meaningful schedule, so also is the development of a product-oriented WBS. A WBS is developed from the SOW, and a trace between the documents through a SOW/WBS matrix helps to ensure that all work is captured in the WBS. A WBS is a management tool that provides project structure and a framework for schedule development and financial management. A WBS Dictionary, or some equivalent document, may also be used to record the definition and content of each WBS element. The structure and format of the schedule should closely correlate to the WBS to ensure traceability and consistency in reporting. This is accomplished by including within the IMS the correct WBS code that associates with each schedule task. In addition to providing a framework for planning, the WBS becomes very important to the P/S by allowing various reporting data to be selected, sorted, and summarized to meet the analysis and forecasting needs of project management. Figure 4-2 provides a sample of a product-oriented WBS with recommended development guidance highlighted.

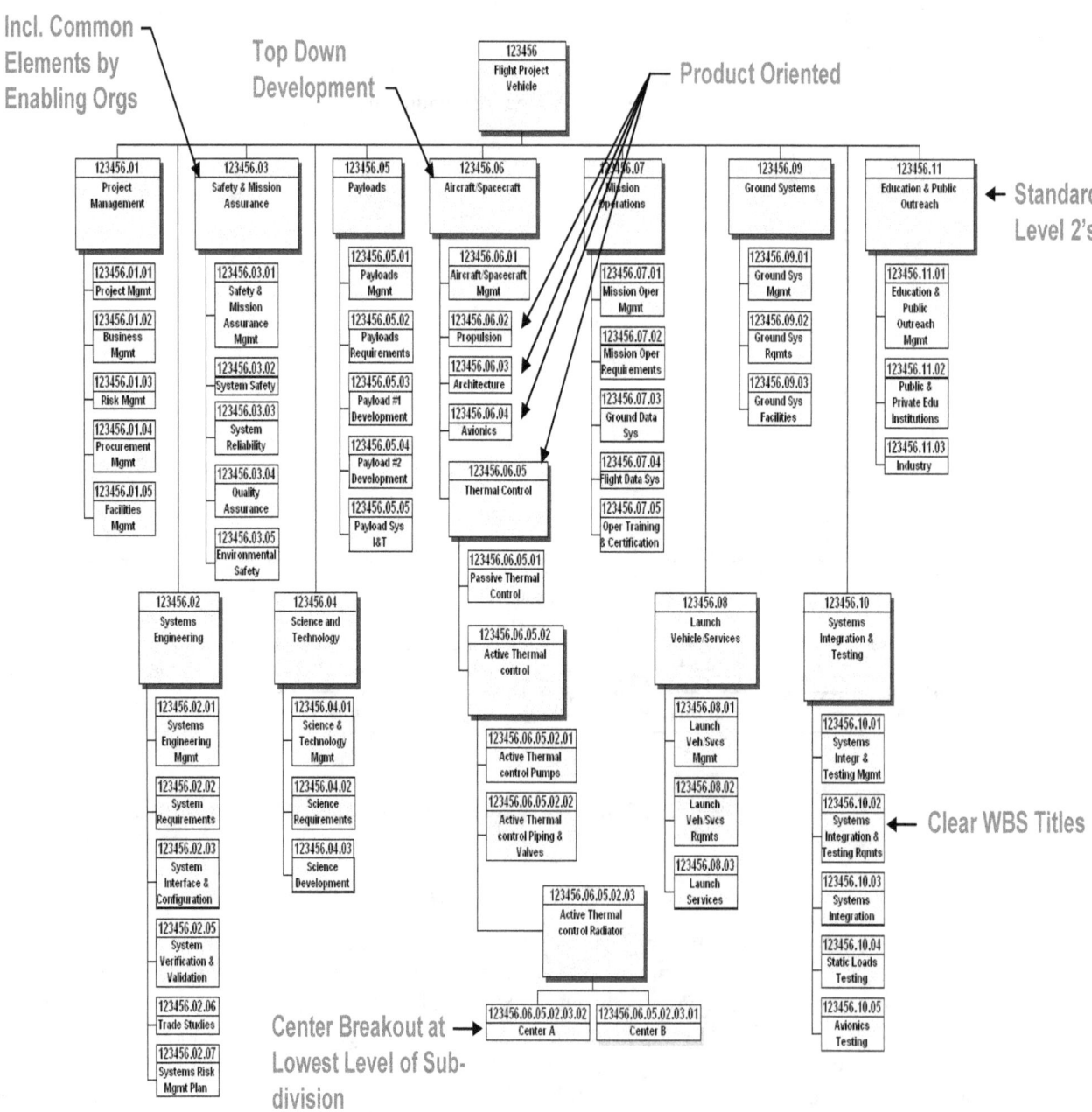

Figure 4-2: Work Breakdown Structure (WBS) Example

A good WBS defines the effort in measurable elements that provide the means for integrating and assessing technical, schedule and cost performance. Since a project WBS plays such a critical role in organizing and managing a project, it is important to know what attributes are involved in a sound WBS document. Below are several key characteristics generally found in a complete and meaningful project WBS document.

- Predominantly product-oriented

- Uses correct standard level two WBS template (from NPR 7120.5 and NPR 7120.8)

- Sub-divided elements are logical, hierarchical, and easy to understand

- Consistent with NASA Structure Management (NSM) coding

- Includes total project scope of work (including contractor effort)

- Allows for work summarization at each level

- Subdivision of work (hierarchy) is aligned with system architecture (e.g., system, subsystem, component)

- Reflects element integration and relationships

The WBS will cover all work elements identified in the approved project SOW, including contract and in-house efforts. Care should be taken to validate that the total project SOW is included in the WBS prior to establishing the baseline for a schedule. The NASA WBS Handbook provides additional examples that can be tailored for most programs and projects.

4.5 Project Organizational Breakdown Structure (OBS)

Many programs/projects require resources from more than one organization or department. The use of a project OBS helps to identify the responsibilities, hierarchy, and interfaces between these organizations. One example of an OBS is shown in Figure 4-3 below. An OBS may be established regardless of whether the organization is structured by function, Integrated Product Teams (IPT), or by matrix assignment. The OBS also identifies the resources available to assign to work activities and to resource load the schedule. When combined with the WBS, the OBS is used to develop a responsibilities assignment matrix (RAM), which clearly identifies which organization is responsible for each task in the schedule (see Figure 4-4). The RAM also helps ensure that no duplication of responsibility occurs. An OBS, if used, should reflect the organizational responsibilities as they pertain to the project. This may differ somewhat, or not at all, from the functional organizational hierarchy.

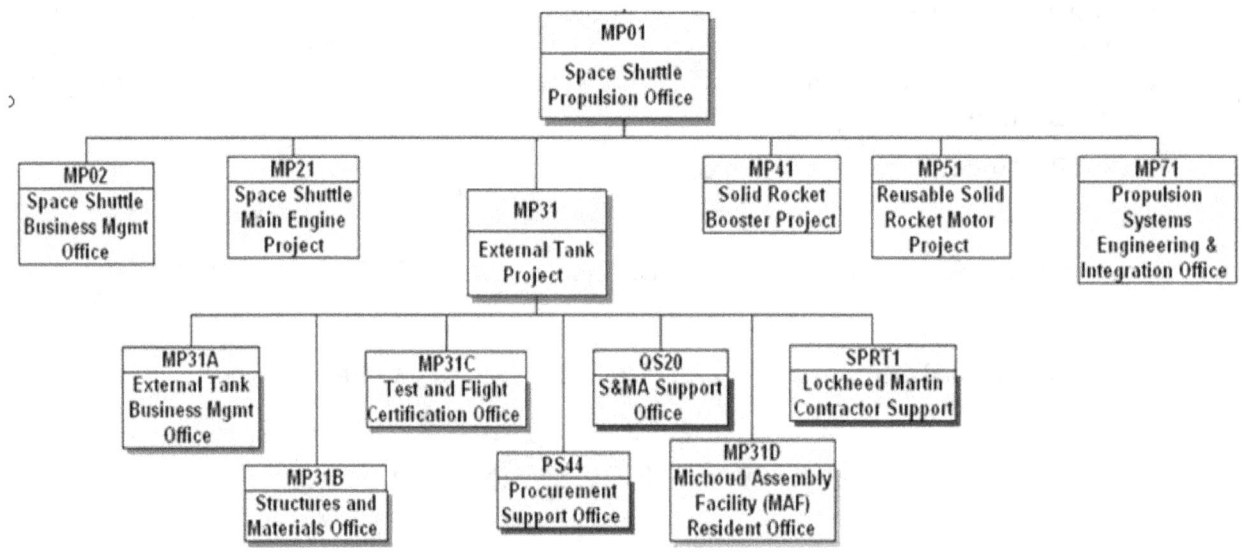

Figure 4-3: Organizational Breakdown Structure (OBS) Example

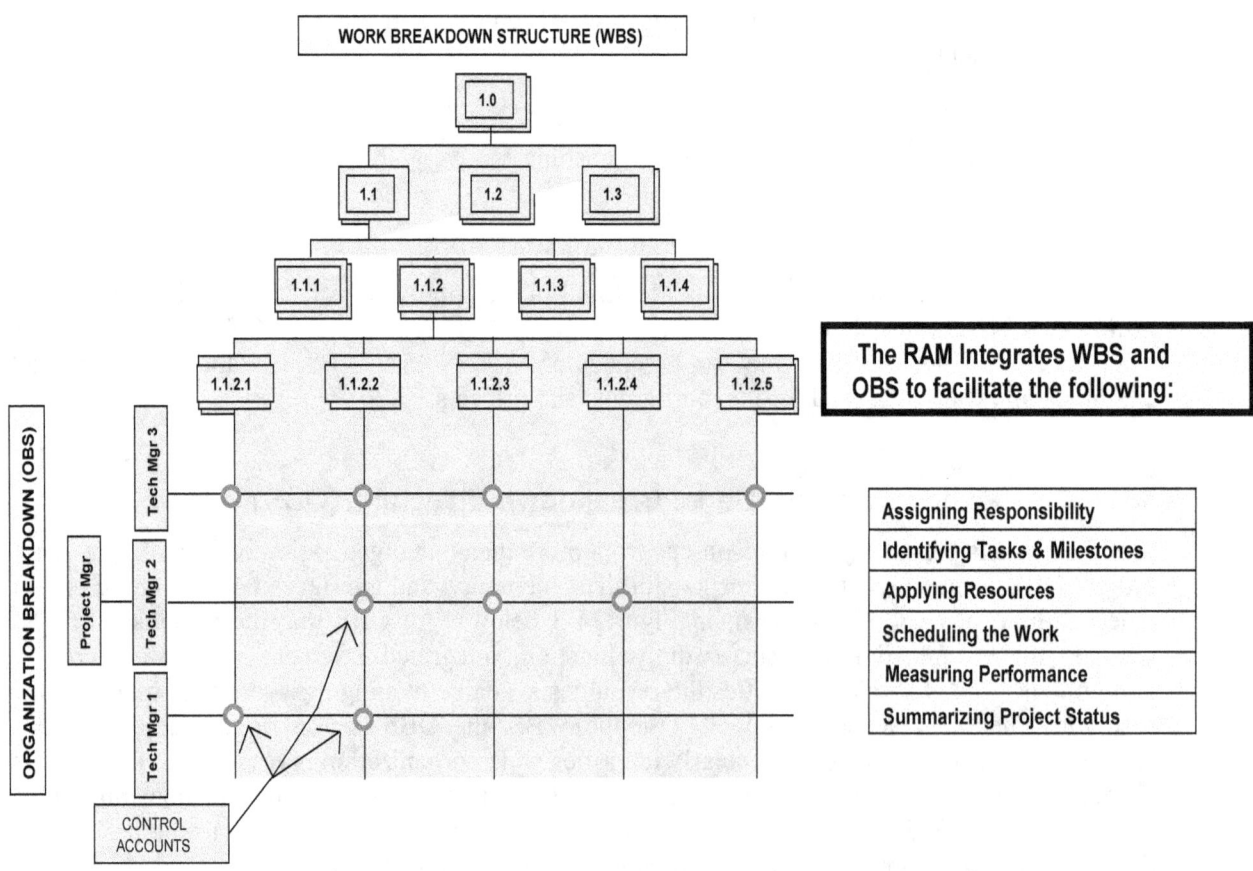

Figure 4-4: Responsibility Assignment Matrix (RAM) Illustration

4.6 Project Funding

From the project perspective, funding must always be a consideration during the schedule development process. And, while the resource loading of schedules may not be an Agency requirement, it is still a recommended planning practice for needed resources and their associated cost rates to be incorporated into the schedule so that the resulting budget estimate can be compared to funding availability and profiles. Any conflicts should be identified and resolved through various methods including the reduction of scope, movement of work, or attainment of additional funding. Funding levels will dictate the amount of work scope that can be accomplished. Typically, funding plans exist, documenting the long- and short-range availability of funds. Budget estimates for planned work, derived from either the resource loaded IMS or other parametric/analogous methods, should also exist. These estimates can provide the P/S insight that will assist in the schedule development process.

Having a clear understanding of the funding that will be available is critical to establishing a credible project schedule. Caution must be exercised in the planning process to ensure the project commitments never exceed authorized project funding. Remember that project funding and the project budget are different entities, but they are related. NASA funding is incremental and refers to the dollars authorized for project expenditure. On the other hand, a project budget plan refers to the value assigned to the time-phased resources necessary to accomplish the scheduled effort. There should always be integration between funding, planned budget, and the associated work content to be scheduled (see figure 4-5).

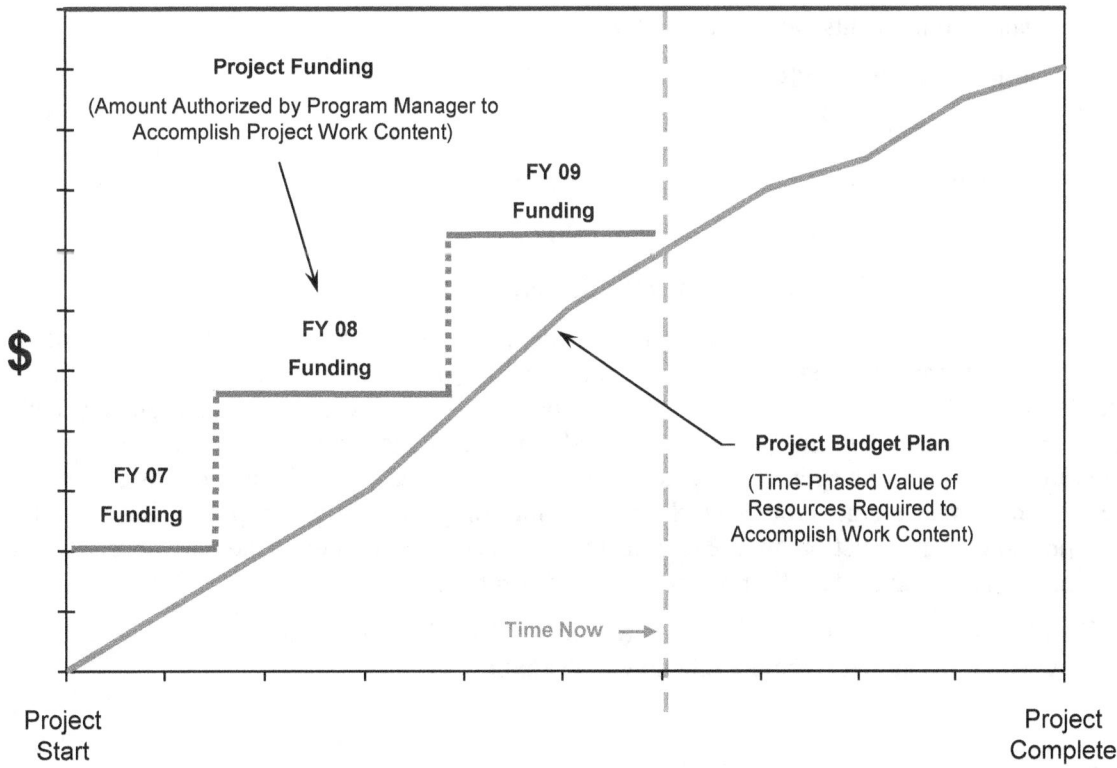

Figure 4-5: Relationship Between Project Funding and Project Budget

4.7 Project Documentation

The P/S should review and use all project documentation that is available to support schedule development. The information gained from these documents provides critical insights needed by the P/S for developing a schedule with a valid basis. These insights may provide key information such as: correct task sequencing, responsibilities, task duration, and resource information. By initially gathering and understanding as much of this data as possible the effort can only lead to a more accurate and meaningful schedule for use in guiding project management.

The following is a list of typical project documents that scheduling personnel should have access to and be very familiar with when starting the schedule development process:

- Implementation Contract
- Project Plan
- Implementation Plan
- Test and Verification Plan
- Scope Definition
- Task Agreements (TA)
- WBS and WBS dictionary
- Data Requirements Document (DRD)
- Project cost estimate/Basis of Estimates (BOE)
- Bill of Materials (BOM)
- Organizational Breakdown Structure (OBS)
- Program Planning Budget Execution (PPBE) data
- Core Financial Business Warehouse Reports

The DRD for project schedules is a critical document for those projects being implemented by prime contractors. Contractor teams build the schedule in a manner that meets the requirements that are detailed in the DRD. If the requirements for schedule format and content are vague or weak, then the contractor's schedule will likely be deficient and potentially unsatisfactory for providing meaningful and accurate data for management decision making. Therefore, whenever possible, it is highly recommended that the government P/S provide input or correction to the project schedule DRD. This will provide clear direction to the contractor project team in their schedule development process. Recommended DRD content for project schedules is reflected in Appendix C.

It should be noted that the information listed in the DRD for Project Schedules is good information to have available for in-house project schedule development as well.

4.8 Baseline Change Log

Changes to the schedule are expected throughout the life of the project. Traceability from the original baseline schedule to the current IMS baseline schedule must be maintained in a disciplined manner. The project manager must establish a schedule change management and control process to handle these changes (see section 8.3). The process and level of detail must be consistent with requirements levied on the project and also should be approved by the Project Manager. A schedule baseline change log to record approved changes to the baseline, as described in Chapter 8, provides a good method for tracking changes and ensures traceability and documentation of changes made to the schedule. Changes to baseline data are important and should be tracked at the level of detail baselined.

There are two general methods for establishing baseline data that in turn influence the methods of baseline control required. The first method is to baseline the entire schedule database. The second method is to baseline a subset of the schedule database, usually key contractual and programmatic milestones. Either method is acceptable. However, the first method (complete baselining) would necessitate a more rigorous, disciplined, and labor intensive baseline control process since any change to schedule baseline data would need to be documented. The second method, sometimes referred to as Control and Notification, still requires baseline control with documented changes but only for those tasks and milestones that have been selected as baseline data. This enables the project team to make adjustments to planned tasks without formal baseline change documentation as long as those changes do not impact the selected baseline data.

4.9 Schedule Requirements

The P/S and project team must ensure that the proper scheduling requirements are included as part of the RFP or contract. A schedule DRD is included in Appendix C. It is important that the DRD reporting requirements be tailored for each project based on only the amount of information necessary to manage the project expeditiously. This can be determined by examining the project's size, complexity and risk. It should be noted that the information listed in a DRD for project schedules is also necessary information to require for in-house project schedule development as well. For in-house activities, a DRD can also be used in agreements with other NASA centers or other government activities. Once again, the DRD should be tailored to meet the needs of each individual program or project.

Chapter 5:   Schedule Development

5.1   Overview

The IMS is developed by defining and sequencing tasks/milestones, estimating task/activity durations, documenting constraints and considering resources. The IMS will contain baseline schedule data, as well as current schedule status and projections.

Schedule management "best practices" help to ensure valid schedule data. Data credibility assumes that all authorized work has been included in the IMS as tasks and milestones, realistic task/activity durations have been utilized, logical interdependencies have been incorporated, and only valid constraints have been assigned. The time-phased sequence of project tasks and milestones is called the Logic Network and should accurately model the approved project implementation plan. This process is called "Critical Path Method" (CPM) scheduling and is considered a standard industry "best practice." When the CPM scheduling technique is carried out in an appropriate automated schedule management tool the result provides accurate current task dates, future task forecasts, total slack values for all tasks and milestones, and the capability for identifying all critical paths for the project. Logic Networks provide a basis for credible time-phasing of all project tasks and milestones. This time-phasing is critical in the implementation of Earned Value Management (EVM) and should be used in the development of the project performance measurement baseline (PMB).

5.2   Data Input and Arrangement

A key consideration for schedule data input is making the data understandable and easy to follow and use. Keep in mind that if the data doesn't make sense it won't be used. Many times it is the simple things we do that make a big difference. For example, making sure that task and milestone descriptions are complete enough to understand what work is being scheduled. This applies to all tasks, not just summary level tasks. While the P/S may know the content of each task in his schedule, that doesn't mean that the project manager, business manager, or even the involved engineers will know it. Additionally, a less obvious, but just as important reason for clear task descriptions is that they expedite and enhance critical path analysis. Identifying and conducting effective critical path analysis entails filtering for only those specific tasks/milestones that are driving the project end date, or another specific milestone, and ordering them chronologically in an understandable waterfall format without the associated summary tasks. A lack of clear understanding of the effort involved in each task makes analysis of the critical path very difficult and time consuming.

Another recommended technique for data input and arrangement is to input the task and milestone data in the order and groupings that make sense. This is not referring to network logic, but just using some organizational forethought when initially entering the data. Doing this reflects a meaningful flow of work, and makes sense even without considering formal logic relationships. This will greatly assist non-P/Ss, and will graphically help those who need to reference or read schedules without having a detailed understanding of the scheduling tool being used.

It is also recommended to input schedule data using a predominantly task-oriented (activities with durations) approach, with milestones included only to reflect major events, key interface points, decision points, and the like. Schedule data that is task-oriented lends itself to a more meaningful approach to monitoring task progress. With this strategy, each activity is reflected in the scheduling tool with an assigned duration that is monitored and updated on a periodic basis to show its progress leading to completion. When tasks are progressed correctly they provide not only a clearer picture of task status, but also an early warning of completion dates that are in jeopardy. If this is the case, the schedule may require new forecast starts and/or completion dates for those specific tasks that are in jeopardy. When only milestones are used to reflect task starts/finishes, it is very difficult to show how the project is progressing toward meeting the milestones. Many times milestones are slipped at the last minute because the actual progress toward milestone completion is not reflected in a clear and obvious manner, and unfortunately, not adequately addressed until it is too late to recover.

A best practice to use for arranging schedule data is coding, which will be discussed later in its own section. Coding enables grouping, sorting, and filtering of data without altering the structure. It is important to maintain the integrity of the data structure while enabling various users with varying needs to query the data effectively and efficiently. Unless required otherwise, it is recommended that schedule data be arranged by WBS. Other techniques that can be used for arranging schedule data are described in the paragraphs below.

Grouping, as used in this section, refers to grouping data together that share some common characteristic. For example, it may be desirable to group together schedule tasks that use the same resource for certain reports. Grouping by values in a resource code field would enable this function.

Sorting refers to ordering data in an arrangement that differs from the natural order as stored in the database. For example, users may find it useful to order tasks by the planned start date. Sorting by planned start in ascending order would generate a list of tasks in order they are scheduled to be worked.

5.3 Task/Activity Definition

5.3.1 Decomposition

Task definition begins with the product oriented WBS. Extending and detailing the WBS down to discrete and measurable tasks is the beginning of the project schedule. Starting with the approved WBS will not only help ensure that the total scope of work is included in the schedule, but also will ensure consistency in the integration of cost and schedule data. The task/activity level of detail should be sufficient to allow for a meaningful measure of progress and the practical establishment of defined finish-to-start network logic relationships.

Task descriptions should be concise yet complete. The task description should be complete enough to stand on its own, but concise enough to facilitate ease of use. Acronyms and abbreviations are acceptable as long as they are standardized and used consistently throughout all project documentation.

A task nomenclature convention or methodology should be established at the beginning of the project and adhered to throughout the project life. For example, a project may elect to use the convention of "noun, adjective, modifier" or "modifier, adjective, noun" for all task descriptions. This type of standardized approach, if used consistently, will make efforts such as searching the schedule database and reporting much easier.

5.3.2 Schedule Detail

The phase and maturity of a project will determine the appropriate amount of detail that can be built into a schedule. In general terms, the earlier in the definition process, the less detail is available (see Figure 5-1).

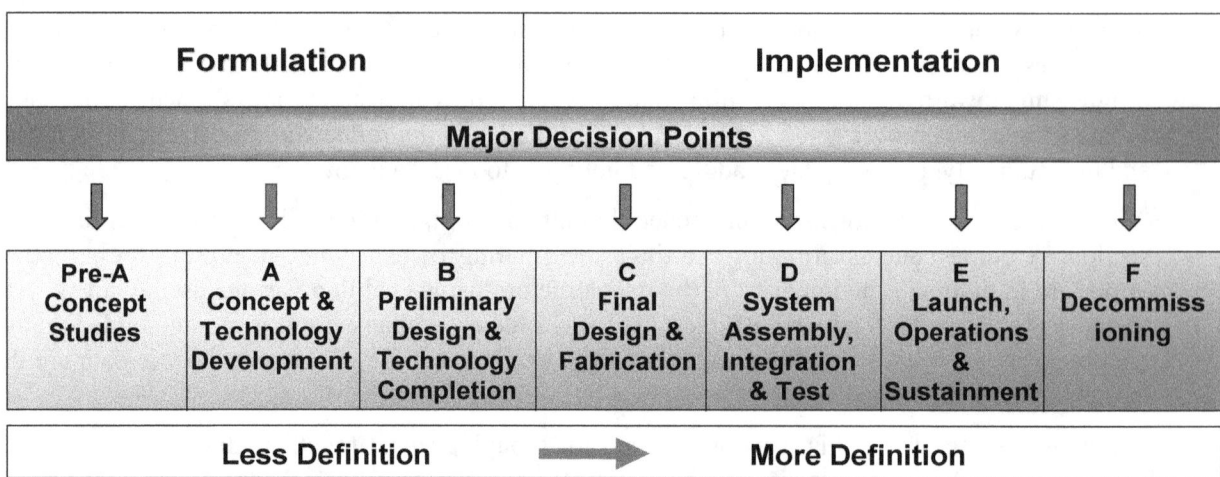

Figure 5-1: Project Phase/Schedule Detail Relationship

The level of IMS detail expected for each incremental development phase within the over-all project life cycle phasing depends on many factors such as: project type (e.g., space flight, IT, Research & Technology), technical complexity, programmatic complexity, technology maturity, project size and dollar value, and whether or not the project is an in-house or contractor implementation, etc. Additional schedule considerations to be addressed within each project phase include: an assurance that all project work scope is included at some level of detail, maintaining an ability to identify the project critical path, and the use of rolling wave (progressive task elaboration) techniques during schedule development and continued maintenance. Regardless of these and other factors or their resulting implications, it is still prudent to establish the following desired expectations for schedule detail associated with each incremental development phase:

- Pre-Phase A, *(Concept Studies)* – A lack of knowledge and understanding of the technology at this early stage, along with immature mission/system requirements, and uncertainty in the industry processes necessary, dictate minimal IMS detail. Typical schedules at this early phase will include major development and integration milestones, such as: major project reviews, major integration points, key hand-offs from outside entities, hardware deliveries, etc. Additionally, it is expected there should be high-level summary tasks reflecting the general time-phasing estimated for developing system/mission requirements, hardware design, fabrication, integration & test, and operational capabilities. These early, high level summary estimates are typically derived from parametric models or historical data from past similar projects. However, it should be noted that detailed information should be available and included at a discrete and measurable level of detail for each concept study that may be involved during this incremental phase.

- Phase A, *(Concept & Technology Development)* – During this phase of Formulation the mission/system concept definition is completed, most concept and trade studies are completed, preliminary requirements are established, and a preliminary Project Plan is developed. Therefore, project definition becomes clear enough during Phase A to allow for a more discrete breakdown of work tasks and milestones. There should be sufficient project definition to accommodate a clearly defined logical sequence of activities leading to product deliverables. Examples of expected detail for a preliminary IMS include, but are not limited to the following: preliminary requirements by subsystem, remaining concept and trade studies, preliminary and critical design by subsystem, long lead procurements, preliminary systems engineering products, preliminary safety and mission assurance products, fabrication flow by subsystem, system/subsystem integration flow, system/subsystem testing, documentation development, flight/simulation software development flow and deliveries, information technology (IT) hardware development and test, test operations development for ground and flight.

- Phase B, *(Preliminary Design & Technology Completion)* – This final incremental phase of Formulation should produce the necessary project definition to allow for discrete and measureable IMS detail, at least for the near-term of nine to twelve months. Near-term effort should be scheduled in meaningful tasks with shorter durations. Durations not exceeding one month are preferable. It is an industry practice to provide IMS task detail down to the level where work is discretely planned and measured at the lowest levels of the WBS and potentially lower where necessary (i.e., subsystem, component, software function, test phase, procurement deliveries, GFE deliveries, interface points, facility modifications, miscellaneous documentation development stages, preliminary orbital debris assessment, etc.) For cases where far-term effort is well defined and task information is already available at the above described low level of detail, then it should also be included in the IMS at the earliest opportunity. Again, the total scope of work should be included in the IMS covering all WBS elements. It should be noted that the project IMS will receive final baseline approval at the end of Phase B (at KDP C), and will serve as the basis for time-phasing of the EVM Performance Measurement Baseline. Task detail should be discrete enough to accommodate the collection of actual time and cost charges at the appropriate WBS level and for the specific work accomplished.

- Phase C, *(Final Design & Fabrication)* – The above Phase B guidance also applies to Phase C. As time proceeds, far-term work tasks with longer durations should be broken down into clearly defined and meaningful tasks with shorter durations (not exceeding one month, or potentially shorter). Special focus should be given to providing clear schedule visibility into the completion of final design by specifying tasks and "release milestones" for specific design drawings or component-level drawings. Fabrication tasks should clearly delineate the necessary work steps that reflect the planned manufacturing work flow. IT development should clearly provide detailed tasks for software functional design, code, debug, unit and integrated testing, software verification and validation, IT hardware development, integration, and test. Specific tasks for Quality Assurance and buy-off should also be clearly identified, as well as, orbital debris assessment baseline documentation. Product delivery milestones from various fabrication process completions should reflect the necessary hand-off points to hardware assembly and systems integration.

- Phase D, *(System Assembly, Integration & Test)* – The above Phase C guidance also applies to Phase D. Again, as time proceeds, far-term work tasks with longer durations should be broken down into clearly defined and meaningful tasks with duration lengths similar to those recommended in Phase C. Special focus should be given to clearly defining the discrete flow of tasks necessary for requirements verification and for hardware and software components to be assembled and then integrated into subassemblies, subsystems, and systems. Schedule detail for this phase should clearly delineate the necessary and measureable work steps that reflect the assembly, integration and test flow of work. Specific tasks for Quality Assurance and buy-off, as-built hardware and software documentation, final systems acceptance reviews, operations procedure finalization, and Operations training and certification should also be clearly identified. Specific hardware deliveries for Launch Operations activities should be included. It should be noted that all pre-launch work should be verified and closed by the Flight Readiness Review (FRR), which precedes KDPE.

- Phase E, *(Launch Operations & Sustainment)* – The above Phase D guidance also applies to Phase E. Operations tasks with longer durations should be broken down into clearly defined and meaningful tasks with shorter durations, as described above. Special focus should be given to clearly defining the discrete flow of tasks necessary for Launch Operations and Sustainment. Activities defined in this incremental phase typically include those tasks necessary for execution of the Mission Operations Plan, such as: final verification and validation reports, flight readiness reviews, final processing of launch hardware, ground operations service preparation for launch, launch activities through achieving operational orientation, on-orbit activities relating to initial and on-going mission tracking, commanding, telemetry, trajectory, systems analysis, mission payload initialization sustainment, and other operational activities as defined in the Project and Mission Operations Plan.
- Phase F, *(Decommissioning)* - This final incremental project phase should also be defined in the same discrete and measurable level of detail as described above. The focus of this incremental phase should address tasks such as: de-orbit preparation and execution, abandonment in-place of flight hardware, recovery of project assets, data and/or equipment disposition and storage, final environmental impact disposition and resolution, lessons learned, contract closeouts, and final public education and notification reporting.

It is important to understand who the schedule stakeholders will be. For example, program managers require less detail for their evaluations than project managers. The level of detail contained in the schedule should also be a reflection of the intended use of the schedule. Management or presentation schedules may contain less detail than schedules used for personnel performing the schedule tasks such as procurement, design, fabrication, or testing. Remember, generally the greater the level of detail in a schedule, the greater the level of fidelity the schedule has. It should also be understood that more detail would normally be required for near-term tasks and also those tasks that are closer to the critical path. It is also highly recommended that high risk and/or high cost areas within the project should reflect more task detail within the IMS. This process will provide better insight to management.

Planning the entire project is essential. One should try to break down a sequence of events so that when one task/activity finishes, another starts. This approach may not always be possible. However, if the schedule is developed using this approach, the effort of logically connecting the tasks/activities will be much easier. Tasks should be detailed enough so that interface points can be clearly identified. These interface points can be where a task is passed from one group to another, a task progresses from one phase to another, or changes in some other significant manner. Additionally, when developing the IMS, remember that every schedule task will eventually be updated. Each task should be clearly and easily identifiable for updating purposes.

Milestones that are important to the program or project should be created. These milestones are not restricted to contractual and programmatic milestones, but contractual and programmatic milestones are included at a minimum, if they exist. There are many other important interface points, phase conclusions, and "hand-offs" where milestones would be appropriate. Milestones should be tied to or represent a specific product or event and should have clear, objective (quantifiable) criteria for measuring accomplishment.

Another factor to consider in determining the level of detail for a schedule is the impact this will have on project Earned Value (EV) and the Earned Value Management (EVM) System. Schedule developer(s) should keep in mind that the level of detail used must lend itself to meaningful cost/schedule integration. It is imperative that the performance measurement baseline (PMB) is based on a schedule plan that is integrated with the budget plan. It should also be noted that the level of schedule detail might determine the type of EV measurement technique (e.g., 0-100, 50-50, weighted milestones, percent complete, level-of-effort) that will be assigned in each earned value Work Package (see Figure 5-2).

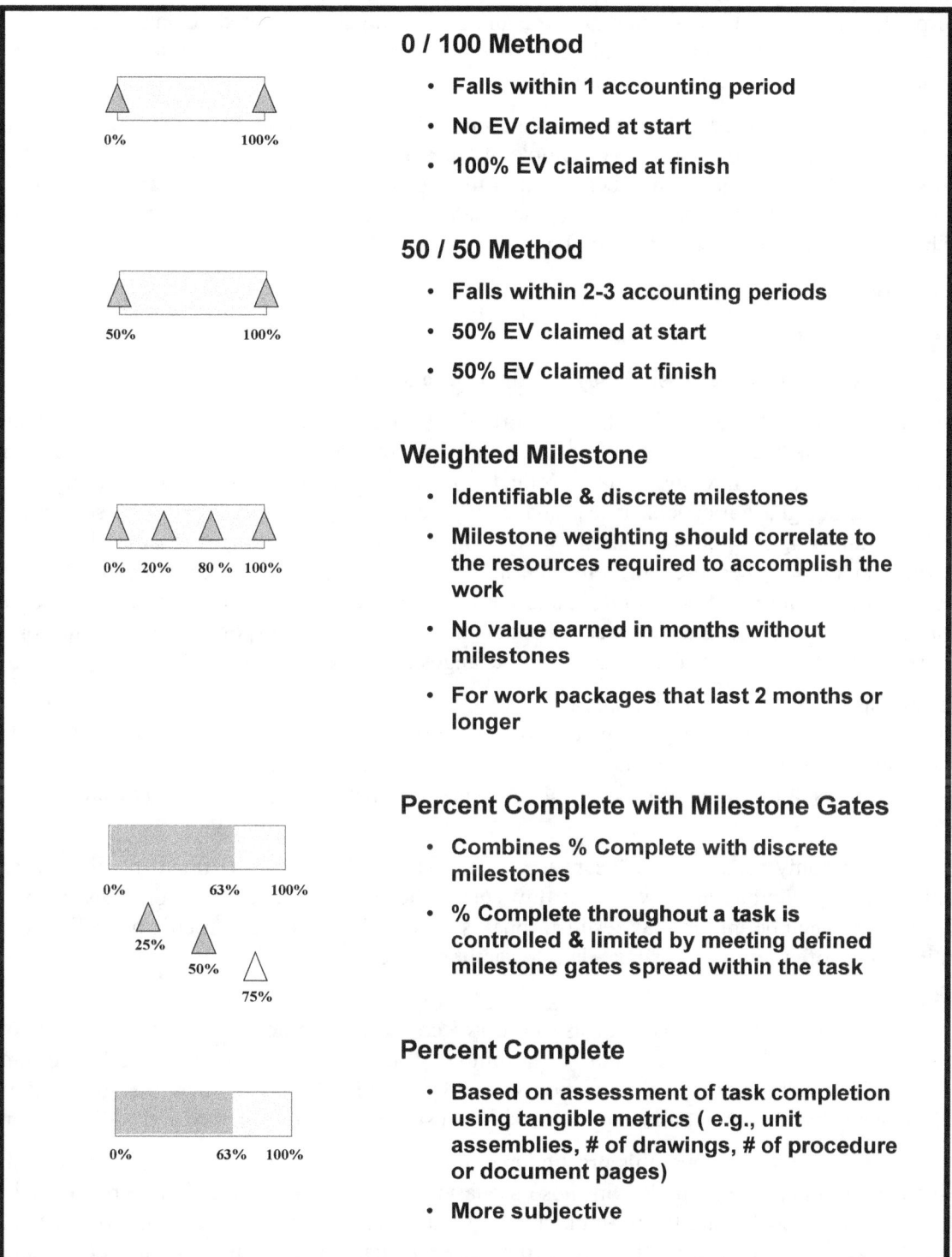

Figure 5-2: Examples of EVM Measurement Methods

5.3.3 Coding

Coding of activities can aid in organizing, displaying and reporting schedule information in useful formats. As noted earlier in paragraph 4.4, the WBS is an important coding structure that should be incorporated into the IMS to aid in extracting and formatting desired schedule data. Appropriate coding can also facilitate consistent vertical schedule integration among master, intermediate and detailed schedules within the context of the overall IMS.

The number of codes needed for a project will vary widely, depending on many factors such as project size, maturity, industry or technology, complexity, entities involved, phase, and so on. The appropriate number of codes to use is the number required to effectively and efficiently manage the project. The same can be said for the types of codes to use. Before deciding whether or not to use a particular code, and how it should be constructed, several questions should be asked and answered, such as:

- For what would this be used?
- Who are the stakeholders?
- Is there another (i.e., easier) way to get the desired results?

P/Ss may occasionally be faced with an opportunity to become creative with regard to coding of data. For example, schedules may need to be constructed that are adaptable to special requirements specific to a particular report or tracking action product. In some cases a request for isolating a particular requirement, design, fabrication, or test phase may be requested. Most scheduling software tools are flexible in allowing field customization for filtering, sorting, and grouping to enable displaying specific criteria. Coding may become more informal in these cases, but should still be documented. It is good practice to maintain a project *coding dictionary*, or some equivalent documentation, to capture code information. In larger programs/projects this document should be incorporated by reference or inclusion in other applicable project documentation with changes controlled appropriately.

The responsibility code is a commonly used code. This type of code can be used in a number of ways, such as a reference to a person, a team, or any other group. This is NOT the same as the OBS, but rather a more specific identifier. The code can be useful for sorting and grouping of data into distinct work groups. The result may then be used to collect status, plan resources, or to communicate a group's work plans.

Another commonly used code is the phase code. This code may be defined in different ways, but usually as a logical grouping of work that flows along the project timeline more or less sequentially. For example, one such definition may result in phases such as Engineering, Procurement, Fabrication, and Testing. It is often used to organize data to facilitate interface-planning efforts and produce summary level reports.

Once a particular code is defined for use in a project, it is recommended that the code value be used consistently for all related project data. For example, if the resource abbreviation "E" for Engineers has been established, this resource abbreviation should be used in all places where a resource abbreviation for Engineers is required. Consistency is the key to a successful data structure and coding scheme.

There may be occasions where this practice may not be practical or possible, due to system limitations or incompatibilities, for example. In these scenarios, a cross reference table can be created to relate pertinent codes. Continuing the example above, let us assume that in our payroll tool the abbreviation "Eng" is used for Engineers. Let us further assume that this is common practice throughout the organization. In this scenario, however, our scheduling tool limits us to only one character for the Engineers resource. We elect to use "E" for Engineers. Our cross reference table would then contain the following entry:

Data Type	Data Item	Payroll Tool	Scheduling Tool
Resources	Engineers	Eng	E

Figure 5-3: Schedule Coding Crosswalk Example

There are an infinite number of codes that either exist or can be created. The tools used along with the project characteristics determine the limitations on this parameter. Other commonly used or customized codes may include task/activity ID, Area, System, Department, Step, Priority, Control Accounts, Flag, Number, Text, and Date codes.

5.3.4 Rolling Wave Planning

The rolling wave concept is a method of scheduling that involves the use of detailed and summary tasks. Long development or repetitive production schedules are examples of project types where this method is typically used.

When using the rolling wave method, near-term tasks are planned to a lower, discrete level of detail. Near-term typically implies 6 months to a year from the current date. Tasks that are scheduled to occur farther into the future may be planned at a more summary level of detail, but still included in the schedule. These summary tasks, while reflecting less detail, should still provide enough definition of future work to allow for identification and tracking of the project critical path that flows to project completion. On a monthly basis, as future summary level tasks come into the near-term window, they should be planned to a greater level of discrete and measurable detail and incorporated into the IMS. It is important to note that while this strategy may be used by many project P/Ss, *it should not used as an excuse for not reflecting the most meaningful level of detail* anywhere in the schedule if the information is already known. It is highly recommended that the P/S plan and schedule in a discrete level of detail as far out as possible throughout all WBS elements in the project. It cannot be stressed enough that planning and scheduling at a discrete level of detail as early as possible will identify and mitigate many project problems, conflicts, and risks.

Again, those who choose to use this method should be aware of the inherent risks. It is quite possible that future detailed planning will reveal situations that, if known earlier in the project, could have resulted in more efficient and less costly work plans. It is a widely accepted theory that advanced planning in the early stages of a project yield significant cost and time benefits when compared to the original cost and time investment.

5.4 Task/Activity Sequencing

5.4.1 Relationships

Logic relationships are critical to accurately modeling a project's planned activities in the IMS. These relationships also provide the means for satisfying the requirement for horizontal traceability within the project schedule. There are four relationship (interdependency) types:

- The <u>finish-to-start</u> relationship - By definition, a preceding activity must finish before a successor activity can start. It is recommended that this relationship be used as often as possible when establishing schedule logic. This relationship provides for the most accurate calculation of total float.

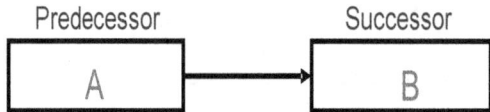

In this example, activity B cannot start until activity A finishes

- The <u>start-to-start</u> relationship - The preceding activity must start before the successor can start. This relationship is used when two activities need to begin at relatively the same time. In most cases this relationship will be used with a lag value. Caution should be taken when using this type relationship in lieu of breaking the effort down into more meaningful and discrete segments of work that can more accurately represent the task sequence. Overuse and/or improper use of start-to-start relationships will potentially hinder true critical path identification.

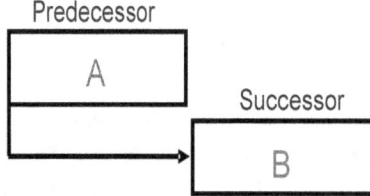

In this example, activity B cannot start until activity A starts

- The <u>finish-to-finish</u> relationship - The predecessor must finish before the successor can finish. This is another relationship that is often coupled with a lag value. This relationship is used when an activity needs to finish and provide something to another activity so that it too can finish. The same caution as noted for start-to-start relationships also applies to the overuse and/or improper use of finish-to-finish relationships.

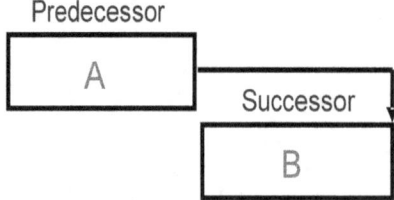

In this example, activity B cannot finish until activity A is finished

- The start-to-finish relationship - The preceding activity must start before the successor can finish. This relationship is almost never used. Caution should be exercised before using this relationship to ensure its use is valid.

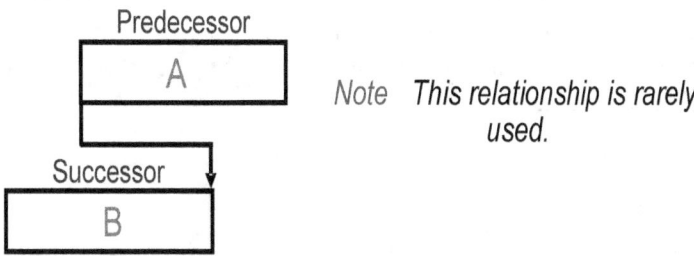

In this example, activity B can not finish until activity A starts

5.4.2 Lag and Lead Use

Lag time is the period of time applied to a relationship between two tasks that delays the defined relationship execution. For example, a task logically tied to another task with a finish-to-start relationship and a 5-day lag will result in the successor task's start being delayed until 5 days after the completion of the predecessor.

A Finish-to-Start relationship with a 5 day lag value means that
Activity B will start 5 days after activity A finishes.

Lead time is the period of time applied to a relationship between two tasks that accelerates the defined relationship execution. The amount of lead time (acceleration time) is assigned as a negative value. For example, a task logically tied to another task with a finish-to-start relationship and a negative 5-day lead will result in the successor task's start beginning 5 days prior to the completion of the predecessor.

A Finish-to Start relationship with a negative 5 day lead value means
the start of activity B will be 5 days before the finish of activity A.

Lead and lag times should only be used when these values represent real situations of needed acceleration or delay time between tasks. Some typical examples where these situations may occur include: cure times on concrete pours, bake-out times for conformal coating of printed circuit boards, and procurement order lead times. Use of these techniques creates a maintenance issue should the basis for the lead or lag time change. Lead and lag times are not very visible and often difficult to discern when analyzing a schedule. They may also corrupt float/slack calculations and incorrectly affect the critical path. There are instances where these types of relationships do exist and are reflected accurately by the correct use of lag and lead times. However, in most cases it would be preferable to use an additional task, appropriately labeled, to represent the lead or lag time and to describe the reason for the lag or lead. This latter practice facilitates visibility and status updates and would likely result in a more accurate and maintainable schedule.

5.4.3 Constraints

A constraint is a fixed date assigned to a task to control when it starts or finishes. Caution should be exercised when using constraints because they are a significant factor in how float (slack) is calculated throughout the project schedule. While it is certainly true that there are various scheduling situations that require the use of constraints to more accurately model the implementation plan, careful thought should be given that they are used appropriately. Common constraint types that can be imposed on an activity include, but are not limited to, the following:

- As Soon As Possible – An Activity or Milestone will finish as early as possible based on its assigned logical relationships and duration. This condition can also be described as the absence of any constraint.

- As Late As Possible* – An Activity or Milestone will finish as late as possible without affecting the schedule end date. ***It is highly recommended that when using "MS Project" this constraint never be used.*** This constraint uses total float to calculate its early finish date instead of free float. This can cause an impact to the critical path and increase the risk for the project end date to slip.

- Start No Earlier Than or Start On or After – An Activity or Milestone will start no earlier than the assigned start date. However, it can start as late as necessary.

- Start No Later Than* or Start On or Before – An Activity or Milestone will start no later than the assigned start date. However it can start as early as necessary.

- Finish No Earlier Than or Finish On or After – An Activity or Milestone will finish no earlier than the assigned finish date. However, it can finish as late as necessary.

- Finish No Later Than* or Finish On or Before – An Activity or Milestone will finish no later than the assigned finish date. However, it can finish as early as necessary. This is a useful constraint to use for a contract deliverable milestone or project completion milestone.

- Must Start On* or Start On or Mandatory Start – An Activity or Milestone will start on the assigned date. Use of this constraint overrides schedule date calculations driven by logic, possibly resulting in a date that is physically impossible to achieve.

- Must Finish On* or Finish On or Mandatory Finish – An Activity or Milestone will finish on the assigned date. Use of this constraint overrides schedule date calculations driven by logic, possibly resulting in a date that is physically impossible to achieve.

- Deadline* - *(MS Project only)* While not listed as a constraint type, a deadline date assignment on any task or milestone has the same results as assigning a "Finish No Later Than" or "Must Finish On." Float (slack) calculations are from the deadline date assignments.

*These types of constraints act as completion points in the schedule, from which the total float value is calculated.

Ideally, *minimal* use of constraints, other than As Soon As Possible, is strongly encouraged. Remember that constraints override task interdependency relationships. Examples where constraints generally have a valid purpose include the following: assigning a "Start No Earlier Than" on a scheduled receivable from an external source, also using a "Finish No Later Than" on the final product deliverable or project completion point. Constraints may also refer to limitations or conditions that affect the schedule. Typical examples of these situations may include test facility downtime or unavailability of specialized computer time/equipment. Take note that for schedules that are resource loaded, these situations are normally best modeled through the use of resource calendars/assignments within the automated scheduling tool. Note that different software tools may have different constraints or even different terminology to describe constraints. For example, at least one popular tool uses the term "constraint" to describe a logic relationship between activities. Other tools have additional constraints such as Zero Total Float and Zero Free Float. While these constraints may be necessary to reflect an actual work situation, **they are the exception and not the rule.**

In summary, while constraint use is sometimes necessary, it should be used only when required to accurately reflect the plan. When used, careful consideration should be given to which constraint type to use. The type of constraint will dictate the impact on float (slack) calculations for the task in question and other tasks logically connected to the task in question.

5.5 Duration Estimating

5.5.1 Determine Task Durations

Prior to estimating durations, a determination should be made as to the unit of measurement and level of accuracy required. For a short term, intense effort, activities may be required to be measured in hours. For a long duration plan, the scheduler may round activities contained in the first year to the nearest day, and in subsequent years to the nearest week or month, then refining the estimates as the activity gets closer. From a schedule analysis perspective, *it is highly recommended to have all duration time units alike within the same schedule.* The reason for this recommendation is that mixing time units within the same schedule may result in slight differences and inconsistencies to slack values internally calculated by the scheduling tool. This result complicates the identification and analysis of a project critical path. Because of this, task durations should generally be assigned in workdays except in cases where more detailed definition in work hours is necessary (e.g., spacecraft vacuum testing).

It is also important to establish whether activities will be measured by elapsed time or by number of working days, and to be consistent throughout the schedule. Alternate calendars can be established to allow activities with different work schedules to be more accurately planned.

Duration is the length of calendar time, as defined by the project, a task is expected to take to complete. Some common methods and sources for deriving or enhancing duration estimates include the following:

- established standards – hourly or daily rates per required quantity based on historical records of accomplishment
- expert experience and judgment - time estimates based on personal knowledge and/or experience with the same personnel, or from similar project work or specialized training
- analogous comparisons using historical or related data
- time estimates based upon historical data gained from past similar or related projects

- parametric analysis - calculated time estimates based upon past data and other related factors such as weight, power, mass, cost, etc.
- team brainstorming – time estimates derived from knowledgeable team discussion and insight
- extrapolations from known data and trends – predicted time estimates calculated from existing known relationships (e.g., 3-point expected value)

It is recommended for P/Ss to always seek valid sources and processes to assist in deriving the most accurate task durations possible for schedule development. It is also recommended that the basis of estimate for task durations within the IMS be documented. This practice will help to ensure schedule credibility is maintained and also provide a critical source for schedule rationale during later project reviews.

Durations should be revisited periodically as work progresses, and as new information becomes available. Durations should not be padded in order to keep a hidden contingency, reduced to be unrealistically optimistic, or arbitrarily cut by management.

If available, make use of historical schedule databases that may exist at each NASA center and/or their supporting contractors.

When interviewing to determine duration estimates, it is a "best practice" for the P/S to ascertain the optimistic duration, the pessimistic duration, and the most likely duration based on duration uncertainty and the probability of associated risks. Calculating an expected value from these three estimates will typically provide a more realistic duration estimate to use in the schedule. The three-point expected value is derived by dividing the sum of the optimistic, most likely, and pessimistic duration estimates by three. Either the most likely estimate or the expected value estimate should be used in the schedule, and the P/S should retain this and other risk information for the probabilistic schedule analysis processes addressed in section 7.2.8. In cases where the duration of a task is very well defined and historically substantiated, the pessimistic, optimistic, and most likely estimates may be the same, or nearly the same value. Examples of this scenario may include the duration estimates for off-the-shelf procurements, standardized testing, and project reviews. In these situations a three-point estimate may not be necessary. A key principle to remember is that the project team member who has the assigned responsibility for a task must also maintain ownership of the schedule for accomplishing that task. This includes their review and approval of the durations contained in the schedule.

Task durations for discrete, near-term effort, as discussed in section 5.3.4 *(Rolling Wave Planning)*, should be kept short and measurable. A recommended best practice for most types of near-term effort is to *not* exceed one month in task duration. Keeping durations to one month or less will certainly benefit projects where EVM is being employed and should result in increased accuracy in performance data. Tasks with durations longer than one month tend to make measurement of objective accomplishment more difficult to assess accurately. This best practice also enhances the P/S's ability to more accurately identify the project critical path.

5.5.2 Schedule Calendars

Resource calendars specify valid time units that a resource may be available to do work. Task calendars specify valid time units that a task or multiple tasks may be worked. Both resource and task/activity calendars should be used where appropriate and be a key consideration when estimating task durations. The P/S should be cognizant of the impact on task scheduling and later schedule analysis when both types of calendars apply. Specific task and resource calendars should be established during initial schedule development. Figure 5-4 reflects a typical scheduling tool screen where task and resource calendars are established for use in the IMS.

Have proper work calendars been set in schedule tool?

- Days / Week
- Hours / Day
- Holidays
- Resource Availability

Note: This is key to correct date calculation!

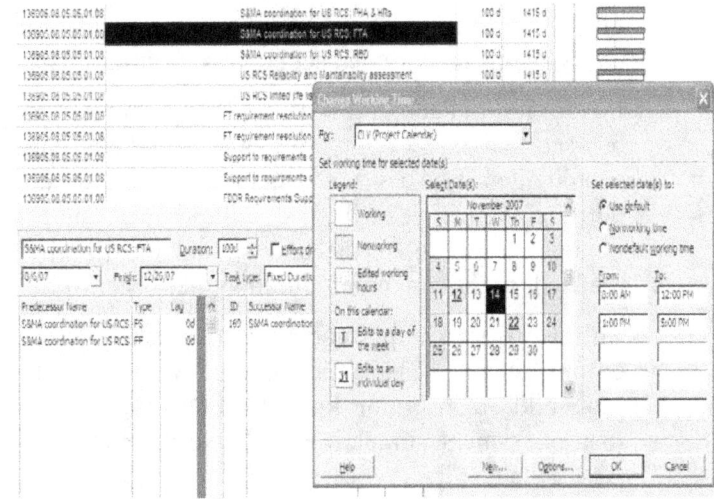

Figure 5-4: Task and Resource Calendar Settings

5.5.3 Resource Impacts

Resources can be classified as belonging to three categories: workforce, equipment, and consumable. In estimating an activity's duration, it is helpful to know what labor resource skills are available and the experience levels of those skills to be assigned. An inexperienced technician or crew, for example, may take longer to perform the task than an experienced technician or crew. While equipment resources are reusable, they may not always be available during the time needed. Consumable resources must be closely monitored and replenished as needed to support schedule needs. All of these factors may not be known at the time of making the initial duration estimate for a task, but they are all considerations that may be used to later adjust a duration estimate, once their impact is known. In addition, labor and financial reports, reflecting actual hours and dollars from prior periods or previous projects, may also provide helpful information for estimating durations. These reports provide historical data that can be used for both initial and re-planning efforts which involve work scope that is similar to previous activities or past projects. The P/S must constantly be vigilant in establishing and maintaining a project schedule that is current and accurate to help mitigate resource problems.

5.5.4 Duration Risks and Uncertainty

Project risks and schedule uncertainty should always be addressed when task duration estimates are made and when the schedule is analyzed. Project risks that impact schedule duration include design issues, fluctuating resources, resource experience, and changes in budget. Schedule uncertainty is due to inaccurate estimates from overestimating or underestimating durations (often referred to as uncertainty), changing or unaddressed scope, task definition changes, and late deliveries.

In addition to each task's most likely duration, an optimistic and pessimistic duration value should also be recorded. The optimistic value should be the least amount of time required to complete the task should identified risks not materialize and/or identified opportunities are realized, or the minimum duration the owner of the task will permit. The pessimistic value should be the greatest amount of time required to complete the task should identified risks materialize and identified opportunities are not realized, or the maximum duration the owner of the task will permit.

This is also an opportune time to assess the likelihood of a task finishing somewhere between the optimistic and pessimistic duration values. If this characteristic is identified, the P/S can assign a probability distribution function (PDF) to each task that will model this likelihood during probabilistic risk assessment simulations. Please refer to section 7.9 for further explanation of the Schedule Risk Assessment (SRA) process.

5.6. Resource Planning

When resources are assigned to tasks within the schedule, this is referred to as resource loading. While there are actually many variations in techniques used for resource loading, this handbook will address only two common approaches that are recommended as part of the schedule management process for a NASA project. The following two strategies should not be viewed as conflicting, but rather as two different methods of resource loading that are used to satisfy different project management needs and purposes.

The first technique is a traditional approach that involves the assignment of specific resources (ie; workforce, materials, equipment, etc.) to the associated tasks within the project's detailed IMS. This approach will be addressed in great detail within this handbook with guidance provided in the paragraphs and figures that follow in this section. Its basic purpose is to provide a tool that yields insight and assistance to the PM and his team in their management of weekly and monthly of resource allocations, and also the on-going evolution of project budget estimates that satisfy various Agency, program, and project budget development requirements.

The second approach utilizes dollars as the only resource and involves the loading of projected costs to associated tasks within a summarized version of the project IMS. Its primary purpose is to provide a management tool that enables the program/project team to conduct a Joint Confidence Level (JCL) assessment. The JCL is a probabilistic assessment that is usually administered prior to key designated program/project life cycle decision points to inform management regarding the likelihood of programmatic success. To clarify further, a JCL will assess the probability that cost will be equal to or less than the targeted cost and schedule will be equal to or less than the targeted schedule date. While it is important to introduce and distinguish the differences between these two common resource loading techniques here, it should be noted that specific guidance and information on this second approach will be addressed in greater detail within other Agency JCL instructional documentation.

Resource loading is not currently an Agency requirement, it is however, an industry recommended practice that provides many additional benefits that greatly enhance the program/project planning and control process. The traditional resource loading approach described above is recommended for use in both in-house NASA project implementations and contracted efforts. While the second approach described above, due to the fact that it is primarily used to enable a JCL assessment by Agency management, will typically be used only by internal NASA management teams.

5.6.1 Identifying, Assigning, and Allocating Resources

Examples of benefits gained from traditional resource loading include, but are not limited to, ensuring accurate integration of work and budget plans, generating accurate inputs for the Agency Program Planning Budget Execution (PPBE) process, providing greater insight into workforce adequacy and allocations, providing cash flow and budget profiles, and providing quicker, more effective analysis for "what-if" exercises. It should be noted that there are also cautions associated with resource loading which may indicate that it should not be done, such as: insufficient team skills for effective implementation and maintenance, inadequate scheduling tools, undefined project resource pool, and a project team culture that is resistant to its use.

As noted earlier, resources can generally be put into three categories: workforce, equipment, and consumables. Workforce resources are the people assigned to do work. Equipment resources are reusable items such as test or manufacturing equipment, and facilities. Consumable resources are resources that have a specified quantity. When that quantity is used up, the resources must be replaced (e.g., fuel, steel, cabling). Resource loading can be done in an automated scheduling tool or in an external spreadsheet. However, to ensure adequate cost and schedule integration, it is generally recommended to implement resource loading within an automated scheduling tool. Prior to assigning resources to IMS tasks, it is recommended that a listing of potential resources (resource pool) be established within the automated schedule tool (see Figure 5-5). The resource pool should contain all types of resources that will be needed for the project, regardless if it is workforce, equipment, or consumables. When developing the resource pool, care must be taken to use a consistent resource naming convention. This will enhance accuracy and consistency in the planning, integration, analysis, and reporting of project resource data. It is recommended that projects not use individual names of personnel as resources due to the size of projects and the need for flexibility to allow multiple people within a single organization to work specific tasks. Once this listing is complete task resource assignments can then be made from the available resource pool listing (see Figure 5-7). All resources required to perform a specific task should be selected and assigned to that task along with quantity required. It is recommended that labor quantities should be loaded in hours. Non-labor resources may be input either as a specific dollar amount that corresponds to each assignment of the specified resource, or as a per use amount that has an associated cost rate that is applied within the automated tool.

Resource Pool: Includes specific planning information for each resource listed

	Resource Name	Initials	Group / Center	Branch	Type	Max. Units	Std. Rate	Accrue At	Calendar
7	ER32 - Combustion Devices	CS	MSFC	ER32	Work	150%	$60.10/h	Prorated	CLV
8	ER32 - ODC	ODC	MSFC	ER32	Material		$10,000.00	Prorated	
9	ER32 - Travel	Travel	MSFC	ER32	Material		$1,000.00	Prorated	
10	ER33 - Component Engineering/Design	CS	MSFC	ER33	Work	50%	$60.10/h	Prorated	CLV
11	ER33 - Travel	Travel	MSFC	ER33	Material		$1,000.00	Prorated	
12	ER33 - Component Test	CS	MSFC	ER33	Work	50%	$60.10/h	Prorated	CLV
13	ER34 - System Design	CS	MSFC	ER34	Work	100%	$60.10/h	Prorated	CLV
14	ER41 - Structural/Stress Analysis	CS	MSFC	ER41	Work	100%	$60.10/h	Prorated	CLV
15	ER42 - CFD Analysis	CS	MSFC	ER42	Work	125%	$60.10/h	Prorated	CLV
16	ER43 - Thermal analysis/modeling	CS	MSFC	ER43	Work	150%	$60.10/h	Prorated	CLV
17	ER43 - ODC	ODC	MSFC	ER43	Material		$10,000.00	Prorated	
18	EV34 - Thermal Analysis & Control	CS	MSFC	EV34	Work	50%	$60.10/h	Prorated	CLV
19	EV41 - Ctrl Sys Des & Analysis	CS	MSFC	EV41	Work	100%	$60.10/h	Prorated	CLV
20	JP30 - ODC	ODC	MSFC	JP30	Material		$10,000.00	Prorated	
21	GRC - Civil Servant	CS - GRC	GRC		Work	100%	$69.33/h	Prorated	CLV
22	JP30 - ESTS	ESTS	MSFC	JP30	Work	100%	$0.00/h	Prorated	CLV
23	ER23 - SDOS	SDOS	MSFC	ER23	Work	150%	$0.00/h	Prorated	CLV

Figure 5-5: Resource Pool Example

Within the resource pool contained in the automated scheduling tool, there are also specific data elements (see Figure 5-6) that must be associated with each resource that are critical to accomplishing effective resource loading. These resource data elements include, but are not limited to, the following:

- Resource name (*employee names are not recommended due to dynamic work assignment changes*, add new resource names as-needed)
- Resource description (e.g., organization name, support contractor company name)
  - Resource Types (e.g., workforce, material, or consumables)
- Element of cost (* indicates recommended minimum):
  - *Travel (designator = Travel)
  - *Personnel Cost (designator = CS)
  - *Other Direct Cost (designator = ODC)
  - *Support Contractor (designator = SUP)
  - Equipment (designator = EQP)
  - Contracts (designator = CON)
  - Material (designator = MAT)
  - Overhead and G&A (designator = OGA)
- Center identifier (use official Center acronym)
- Maximum number of units available Standard Unit Rate (project to determine)

- Overtime Rate (project to determine)
- Cost Per Use (project to determine)
- Accrual method (start, prorated, end)
- Resource Calendar (reflects active periods of resource availability - project to determine)

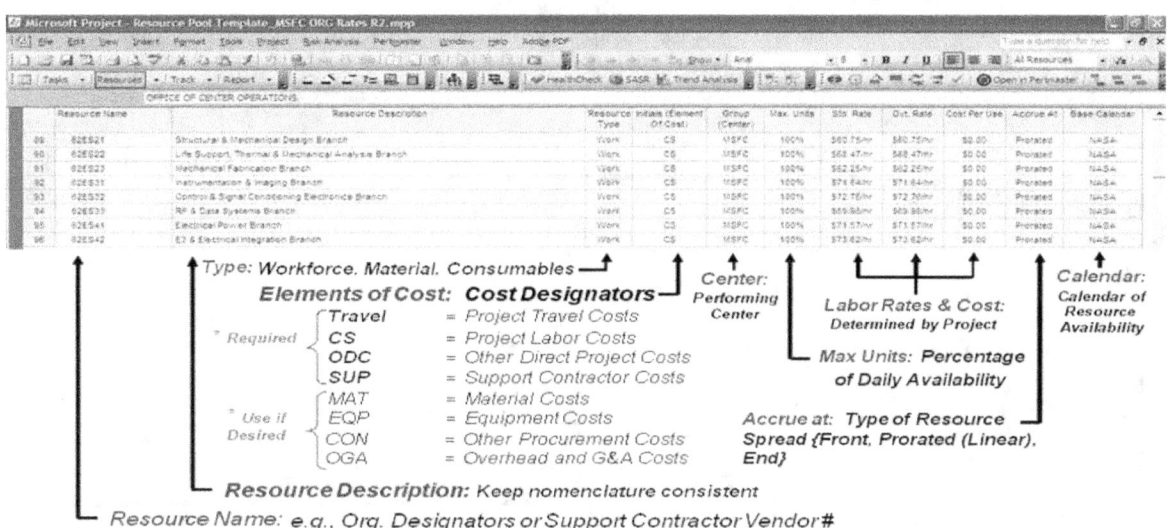

Figure 5-6: Resource Pool Associated Data Elements

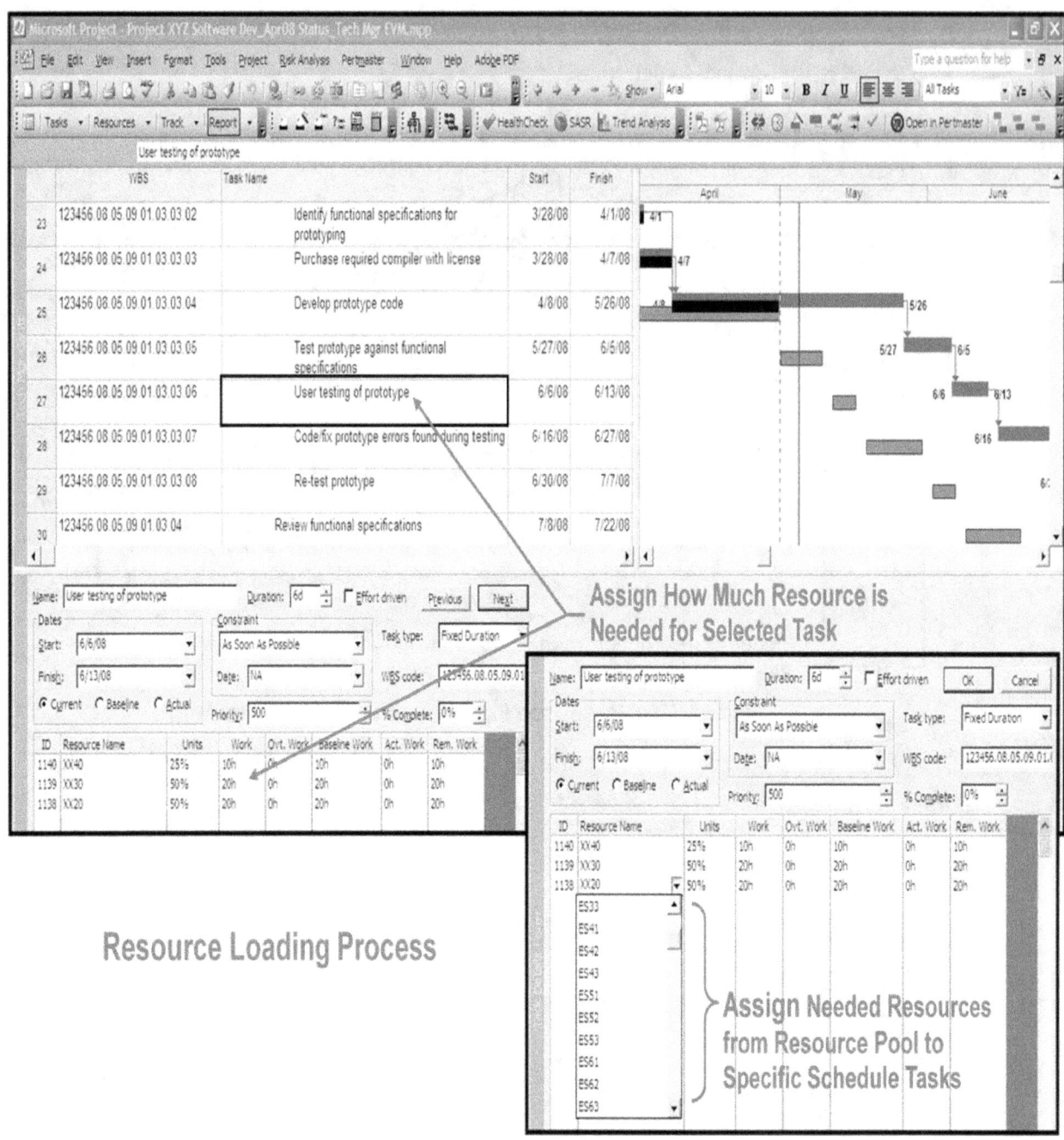

Figure 5-7: Schedule Resource Loading

Allocation of resources may be done in various units of measure, depending on the type of resources used. The P/S must also ensure that resources are distributed adequately across the specified task durations. Most automated scheduling tools distribute assigned resources in a linear fashion, evenly across the duration of a task, unless the user takes action to customize this distribution. To ensure that a reasonable and achievable schedule plan has been developed, it is important that the P/S diligently work through the resource loading process and establish a complete and credible basis from which to move forward during project implementation.

Resource loading can provide valuable information regarding over or under allocations in a schedule. This information can alert the project manager if a task cannot be completed in the time scheduled due to a resource shortage or if adding more resources can shorten the task duration.

5.6.2 Resource Leveling

Resource leveling is the process of moving schedule tasks without violating network logic or constraints in order to achieve a more consistent level of resources throughout the schedule duration. Analysis through resource leveling is a process that can only be accomplished when a schedule is resource loaded and should generally be left to the more experienced P/S. Additionally, for the process to provide credible data, the schedule must be structured as an end-to-end logic network with all interdependencies identified. Resource leveling is generally accomplished through the use of a proven automated scheduling tool that has the capability of electronically evaluating total float values, logic relationships, constraints, and the amount of resources applied to each task or milestone in the schedule. In carrying out this function, the management tool will reschedule tasks as allowed, based on the characteristics listed above, to most efficiently utilize and level the number of resources in an effort to eliminate over-allocations. Visual examples that illustrate the results of resource leveling are provided in Chapter 7 of this document (see Figures 7-12 and 7-13). It should be noted that the resulting schedule data should be reviewed by the management team carefully, and not just taken at face value, to ensure credibility for project implementation. Other concerns may exist relative to schedule data that are not necessarily related to float (slack), logic, or constraints that may require adjustments to be made before baselining the schedule.

Resource loading and leveling is the recommended method to accurately validate whether the scheduled project completion is achievable with the allotted resources available. To baseline a project schedule without first resource loading and conducting leveling analysis is to assume a significant risk in achieving project completion within budget and on schedule.

5.6.3 Resource Rates

Rates for each resource are determined and applied to calculate the cost of labor, material, or other resources assigned to each task. Resource rates are typically maintained within an automated schedule tool and applied to task resources contained in the IMS. Resource rates may also be maintained and applied in a separate cost management tool that may be part of the project management process. Applying resource rates to task resources is a recommended bottoms-up method of determining the time-phased budget baseline for a project. This process facilitates the implementation of earned value management (EVM) within the project. This process is also useful in facilitating evaluations of cost impacts due to schedule changes.

5.7 Schedule Margin Planning

It is a recommended practice that schedule margin, based on risks, duration uncertainty, and historical norms, be clearly identifiable when included within the IMS. Schedule margin may also be referred to as "schedule reserve" or "schedule contingency." Schedule margin is owned and controlled by the Program/Project Manager. Factors that may contribute to determining the amount of schedule margin are: a) expert judgment, b) rules of thumb, c) % of overall project (or activity) duration, d) calculated by expected value of risk impacts, or e) through insight gained from a probabilistic schedule risk assessment. For clarification, it should be understood that schedule float (slack), which is a calculated value based on network logic, should not be considered as schedule margin.

Schedule margin is used for future situations that are impossible to predict (for unknown unknowns). For this reason, no specified budget should be assigned to a schedule margin task because there is no known project scope involved. However, at the same time, it is very important that an adequate amount of project management reserve be available to cover a reasonable workforce level through the duration of the schedule margin activity. Schedule margin is a separately planned quantity of time above the planned duration estimate reflected in the IMS. Schedule margin is intended to reduce the impact of missing overall schedule objectives. The preferred technique for including schedule margin in the IMS is to insert additional tasks that are specifically identified as "Schedule Margin." These tasks should have durations assigned that provide the additional quantity of time deemed necessary to absorb the impacts of unknown schedule risks. Schedule margin must be inserted into the IMS at strategic locations so that it satisfies its intended purpose as overall schedule management margin for the project completion. To ensure this, it is recommended that this type margin be placed at the end of the IMS network logic flow just prior to hardware delivery or whatever the appropriate project completion task/milestone might be. Other example locations for this type of margin might include placement just prior to PDR and CDR. Only by clearly identifying the amount and location of schedule margin within the schedule can its use be tracked and managed adequately.

An alternative technique for managing schedule margin involves the use of milestones, constraint dates, and relationship lag values. For example, suppose there is a programmatic or contractual event commitment of concern (typically the project completion point, hardware delivery, launch, etc.). This event may be entered as a milestone in the IMS with a "Finish No Later Than" (FNLT) constraint which specifies the hard commitment date. Another companion milestone, which references the same event, may be entered as a predecessor to this event with no constraint and is labeled as the "target" for the event. The interdependency relationship connecting the two milestones should also include a specified lag value. The specified lag value represents the amount of schedule margin that is considered necessary by the project team. The schedule margin represented by this lag value is to be managed and controlled by the Project Manager. When using this schedule margin technique it is very important that the P/S monitor the lag value along with the amount of slack on the associated "target" milestone at every schedule update to assess and control the use of schedule margin.

When incorporating schedule margin into the IMS, there are key guidelines that should always be addressed and maintained throughout program/project implementation. First, schedule margin should always be identifiable in the schedule. Second, schedule margin should be managed and controlled by the Program/Project Manager. Third, since no budget should be assigned to schedule margin, an adequate amount of management reserve must be available to cover this added duration. It is clear that schedule margin that cannot be identified probably cannot be controlled. Caution should be taken when using the alternate technique described above in providing necessary schedule guidance and/or assumptions that clearly outlines to the community of schedule users just how schedule margin is identified in the IMS. Caution should also be exercised, regardless of the technique used, to have an established process in-place that ensures the Program/Project Manager has both the necessary insight into how much schedule margin is available, and also the capability to control its use.

5.8   Establishing the IMS Baseline
5.8.1   SOW/WBS/Schedule Correlation

Prior to IMS baselining, it is essential to validate that SOW, WBS, and schedule correlation exists. The SOW document should contain a narrative description of the entire scope of work for the program/project effort in a structured, product-oriented fashion. The WBS should also reflect the entire scope of work for the program/project in a structured and even more defined product-oriented fashion.

Each of the WBS elements should be reflected in the project schedule. Furthermore, each WBS element should have at least one corresponding schedule task. Therefore, each scheduled task/milestone should reflect the related WBS element in the appropriate data field.

It is equally important to ensure that each schedule task/activity has a WBS element that matches a WBS element in the WBS dictionary. Any schedule task/activity that has a WBS element not contained in the WBS dictionary may reflect either work not in the project scope or not labeled correctly.

5.8.2 Schedule Accuracy and Credibility

Network logic should be reviewed by the P/S to ensure that it is complete, accurate, and realistic. Within the IMS, there should generally be a minimal number of tasks with no successors or predecessors identified, and these cases should only occur with valid reasons which are documented. Forced or fixed dates (constraints) should only be used when network logic cannot accurately depict the true sequence of work because of some external influence or an influence beyond the control of the project team members. The constraint types should be reviewed carefully for accuracy and desired effect. "As Late As Possible" constraints should generally not be used in an IMS. Improper and/or invalid use of constraints should be minimized due to the potential for creating misleading schedule data in date and float (slack) fields.

A thorough effort should be made to identify the tasks that may be worked in parallel with other tasks, tasks that must be worked in series with other tasks, and also tasks that may be worked once another task has progressed beyond a given point. Each of these situations can be reflected with the proper use of logic relationships and lag or lead values.

Once these details are verified to be correct, the P/S should examine the schedule from the time-phased sequencing perspective. Tasks/milestones should be arranged in a logical sequence. For example, the design for a component would normally precede its procurement, which would precede its delivery, which would precede its installation and testing, and so on. Another example would be that the majority of design tasks/milestones should be completed before a final or critical design review milestone. No task/activity or milestone that is required to be completed in order to finish the project should occur after the project completion milestone.

A network logic review by project team members should focus on three specific areas. First, within each logical grouping of work, the sequence of tasks/milestones should be verified. This may involve team members from different organizations or multiple project personnel from the same organization or both. Second, each interface or "hand-off" between different work groups should be verified. And finally, the overall project phasing sequence should be validated.

The P/S should verify that the duration for each task/activity entered is accurate and realistic, based on the information provided for that task/activity. The method of verification is dependent upon the credibility of the source of the original duration information. All assumptions made in determining task/activity durations should be recorded. This can be an especially important consideration when later assigning resources to scheduled tasks/activities. Knowing the basis for task durations may also aid in setting risk parameters when conducting schedule risk assessments.

With each schedule task now relatively well defined, the duration for each should be verified with the task owner. All changes have to be evaluated for the impact on other related or logically tied tasks/milestones. All specific assumptions that are part of the basis for determining the duration of a task should be recorded. This should include the impact on the duration due to the experience or skill level of the resources to be assigned to each task.

5.8.3 Float (Slack)

There are two types of float (slack) common to most scheduling tools. They are "total float" and "free float." Total float is defined as the amount of time that a task or milestone can slip before affecting the project end date. Free float is defined as the amount of time a task or milestone may move into the future from its early finish date before affecting its immediate successor task(s). Schedule analysis utilizing both types of float values is common, but each must be used for its appropriate application. It should be noted that accurate float values can only be determined if a complete and validated network logic is in place. While free float is valuable in analyzing scenarios involving schedule impacts and conflicts for a specific set of tasks/milestones and prioritization of resource utilization, it should generally not be used as a means of monitoring and managing schedule performance for the total project. An important key to achieving the desired schedule completion date is being able to identify and evaluate what tasks are directly driving the project end date. Total float provides this capability, and knowing the total float for every task and milestone in the schedule will provide management with the necessary insight into how each task impacts the project end date. Credible schedule management is not possible without proper analysis of float values.

5.8.4 Schedule Risk and Uncertainty

A schedule risk assessment (SRA) is crucial during project formulation and implementation planning. It is an important analysis process used to assess potential schedule risks, duration uncertainty, and to evaluate the likelihood of a schedule being achievable. This process will also provide completion probabilities for key events identified in the schedule. Although there are various ways of conducting a schedule risk assessment, the recommended technique is through the use of a proven risk analysis tool with the capability of computing simulations based upon realistic risk parameters assigned to each task/activity within the schedule. The risk parameters associated with tasks/activities include the minimum, maximum, and most likely durations expected for each task, as well as a distribution profile. The distribution profile, also called the probability distribution function (PDF), models the likelihood of task/activity durations between the minimum and maximum values.

As numerous simulations are executed and calculated, the risk tool will factor in the assigned likelihood parameters and distribution profiles, and utilize the assigned task/activity interdependencies to provide probability percentages for achieving key selected events within the schedule. These percentages will aid the management team in determining an adequate amount of schedule margin to be included in the schedule before baselining.

It is recommended that the P/S collaborate with the responsible technical managers in gaining as much information as possible prior to assigning the required risk parameters. This collaboration process will aid in ensuring that the assigned parameters are realistic and lead to more reliable schedule risk data for making management decisions. During this process it is also helpful to use the identified risks from the project's risk management system and relate them to specific tasks/activities within the schedule prior to assigning parameters. This collaboration will lead to a more accurate schedule risk assessment prior to baselining. For more information on conducting an SRA, please refer to Chapter 7, "Schedule Assessment and Analysis."

5.8.5 Baseline Approval

Establishing a schedule baseline is a very significant step to take when putting into place the necessary tools for managing project performance. NPR 7120.5 states that the project IMS is part of a project's baseline content. A preliminary IMS baseline is to be established for KDP B then followed by a formal, project-approved IMS baseline to be in place at KDP C. It is crucial that the schedule baseline encompasses the total approved technical scope of work and accurately models the project team's plan for implementation. The project schedule should also be integrated with the project's budget plan. The IMS baseline should never be approved unless the schedule plan consistently aligns with the budget plan. Review and approval of the schedule baseline should not be taken lightly and changes to it should be controlled carefully. This is especially true when utilizing earned value management techniques. The approved schedule baseline becomes a key component in the project's Performance Measurement Baseline (PMB), and as such, becomes a key part of the yardstick by which all project performance is measured and analyzed.

Affected resource managers (RM) or supervisors should participate in this review, whether directly or indirectly. Ultimately, the project should have a commitment from each affected RM to provide the required resources in the timeframe indicated by the approved baseline schedule.

One very simple, but useful technique for analyzing project resources during schedule development prior to baselining is a Summary Level Cost/Schedule Correlation Check (see Figure 5-8). From a high-level perspective, this tool will provide insight that should be helpful in determining if the proposed resource plan is consistent with the proposed schedule plan. This simple technique will provide a means for quickly determining if planned resource usage peaks correlate appropriately with major project milestones and phases.

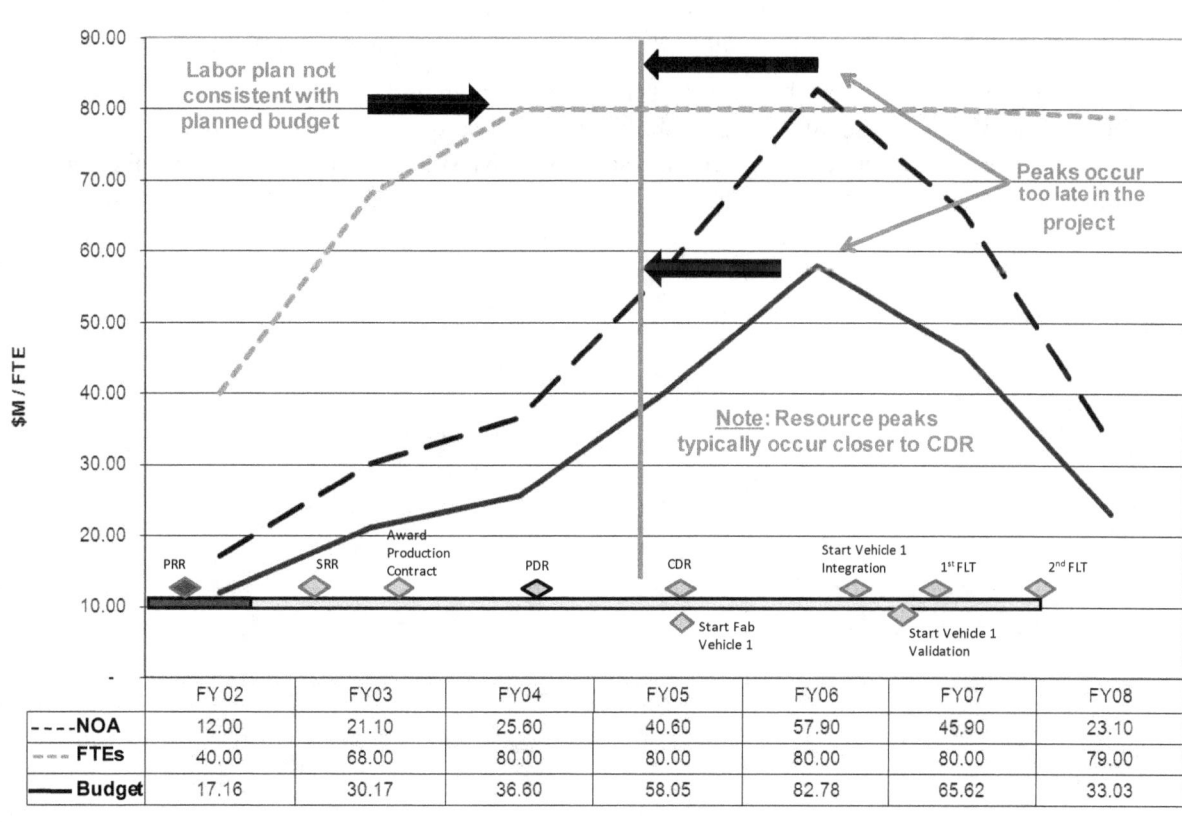

Figure 5-8: Summary Level Cost/Schedule Correlation Check Example

As data is reviewed and changes are identified, the final revisions should be made to the schedule. At a minimum, the P/S should review the results of these changes with all the affected project team members. Major project or program milestones that have changed should be verified and approved by the project or program manager and the affected organizations. In the event that the result of these revisions causes a major milestone date change that is unacceptable, the project team must return to the review process.

The final review should also ensure that there is adequate schedule margin and that it is clearly identified as such in the project schedule. One method of ensuring this is to perform a schedule risk assessment (SRA) prior to approving the baseline. Another typical method is to identify project risks that may impact the schedule and estimate the impact duration for each risk. The summation of all risk impact durations will provide an overall estimate of schedule reserve needed. These techniques are highly recommended.

All stakeholders affected and responsible entities should commit to project management to adhere to the plan as reflected in the baseline. This will involve all required resource and task performance (i.e., dates) commitments. The baseline should include all work in the project and reflect the actual plan for executing the work, in both sequence and time frame.

Once accepted and approved, an electronic copy of the project baseline should be saved. Schedules for contracted work should be subject to configuration management. This baseline should be used for comparison throughout the life of the project. All proposed changes to the baseline must be evaluated carefully to determine cost, schedule, and technical impacts, and then approved only by the appropriate project management designee. IMS milestones that are controlled by the program office may only be changed after receiving approval from the program manager. The current IMS should begin as a copy of the approved baseline and reflect changes made to the plan by the program/project team.

The baseline schedule should also reflect the resource plan and budget plan for the project. All three of these (schedule, resource, and budget) should be in phase with each other. This baseline will be used to establish the Performance Measurement Baseline (PMB), which in turn will be the basis for earned value calculations and analysis later.

Chapter 6: Status Updates and Schedule Maintenance

6.1 Overview

As project work is executed, all tasks/milestones in the schedule should be updated to reflect their current status. This will involve timely updates to network logic, task percent completes, resource allocations, remaining durations, and actual start and finish dates. It is critically important for the IMS to accurately and realistically reflect the current plan to complete the remaining authorized scope as contained in the baseline. To ensure accurate project analyses, it is important that all incomplete tasks/milestones in the schedule be updated to a single status date, including tasks/milestones that should have started or completed, but haven't.

6.2 Status Update Accounting

6.2.1 Update Methodology

For projects not requiring EVM discipline, the IMS update process may be carried out with a little more flexibility in the techniques used. However, particular attention and discipline are still required by each P/S during schedule updates to ensure that the IMS continues to reflect an accurate model of what has been accomplished and also of the on-going implementation plan. The following guidance is provided to address the typical process flow encountered during a routine IMS update:

- Copy and archive IMS versions prior to each update cycle. This will ensure proper historical records for future audit activity and also to provide a source of reliable schedule duration information for future duration estimating and validation.

- Gathering task/milestone status may be accomplished in various ways such as providing task owners with a printout containing their specific tasks that require update information, face-to-face meetings with task owners to discuss and redline the schedule copy, or establish weekly, bi-weekly, and/or monthly project IMS update meetings with all task owners participating by verbally providing their status. Regardless of the strategy for gathering updates, the P/S must ensure that progress given is consistent with the pre-established task completion criteria.

- Incorporate the gathered status updates into the IMS. It is important to understand that many scheduling tools offer different "% complete" fields with different functions. The "% complete" commonly in the default view for most scheduling tools is actually "percent duration complete" and may or may not be directly related to the "physical work percent complete." This is particularly important when using earned value.

- Do not leave any task or milestone un-progressed and time phased to the left of the current status date. Reflect tasks as actually started and in-progress (with proper completion forecast), actually completed, or re-forecasted to a more accurate start and/or completion date. *(Beware that some schedule management tools, such as MS Project, do not force the user to update the status of on-going or behind schedule tasks/milestones).*

- Use "Remaining Duration" as the primary method of statusing in-progress tasks. This will keep projected finish dates accurate, as well as succeeding linked tasks/milestones properly time phased.

- After all status updates have been incorporated into the IMS, it is important to analyze schedule impacts and resolve all issues resulting from the new status updates. This analysis includes, but is not limited to: identifying the current critical path and comparing to the previous critical path, identifying and correcting status input errors, identifying tasks/milestones with missing status, identifying new schedule related risks, identifying necessary logic, resource, and calendar changes that are required, etc.

- After status updates are incorporated and impact analysis and resolution is complete, provide updated schedule reporting to project management and necessary customers.

For those projects requiring EVM discipline, the following update processes should also be carried out in conjunction with the above update process guidance. The status of work packages is derived directly from the objectively determined status of the time-phased tasks/milestones composing the work packages. Work packages containing deliverable products, or work associated with deliverable products, are deemed "discrete effort." Discrete effort work packages are assigned an appropriate Performance Measurement Technique (PMT), considering duration, value, and nature of the effort. PMT may be categorized in one of three ways: as weighted milestone, percent complete, or apportioned types.

- Weighted milestone – significant events are represented by a milestone that is assigned a percentage of the total value of the task/activity.

- Percent complete – is either an objective (e.g., based on physical quantities) or subjective (personal judgment) determination of the percent of the task/activity that has been completed. It is strongly recommended that as each task percent complete is determined and incorporated that the task's remaining duration is also determined and accurately reflected in the IMS. It should be noted that when updating an in-progress schedule task, it is the remaining duration that becomes the determining factor in reflecting the task's accurate forecasted completion date.

- Apportioned – is determined to be the same percent complete as the related task or tasks (e.g., safety/quality inspector support for fabrication of hardware).

PMTs are individually selected for each work package to enable the most accurate assessment of performance possible. Future activities, requiring further definition, are assigned to planning packages and are reflected in the IMS at a summary level of detail. As planning package tasks reach the near-term window, they are divided into discrete work packages, and assigned appropriate PMT, prior to beginning work.

Tasks that represent only support efforts (e.g., project management, administration, safety) are typically referred to as "level of effort" (LOE) tasks. These tasks generally have no discrete products that are produced making the quantification of accomplishment difficult or impossible. Because of this characteristic, LOE tasks are measured with the passage of time using a percent complete measurement technique based on the baseline duration of the task. Due to the nature of LOE tasks, they should never reflect a schedule variance.

The same PMT method used for planning purposes should also be used for claiming earned value.

The project IMS should at all times reflect as accurately as possible the current plan for accomplishing the remaining work. This will involve updates to network logic, remaining durations, and actual start and finish dates.

6.2.2 Update Frequency

Status updates should be made as frequently as feasible. The frequency many times is dependent upon what phase the project is in, who is doing the work (in-house NASA, contractor, or both), as well as the number of resources available to gather, input, and analyze the new status updates. Typically, early in a multi-year project, a monthly update is adequate, but if the necessary resources and processes are in place, then weekly or bi-weekly may be the preferable interval. Project scope that is being implemented by in-house NASA organizations provides the flexibility of updating progress weekly or bi-weekly. Projects with prime contractor involvement will generally get an updated IMS from the contractor only on a monthly basis. This situation limits the capability of providing management with a fully integrated and updated IMS to a monthly cycle. There are always some exceptions to the update guidelines established by a project that may come into play, such as, the type of work being done may dictate the frequency or the level of visibility required. Also, some schedule items are designated as "management reporting" tasks/milestones and may require a more frequent update cycle.

6.3 Schedule Maintenance

Schedule maintenance should consist of, but not be limited to, verifying/modifying task/milestone durations, revising or adding logic interdependencies, additions of new tasks or milestones and their associated logic ties, re-allocation or assignment of resources, calendar changes, and other miscellaneous minor schedule adjustments. Schedule maintenance is an on-going process that is typically carried out in parallel with incorporating status updates; however, it should be done as IMS modifications are necessary apart from the normal status update cycle. If scheduling software being used by the project does not enforce sound update discipline, then at every status interval, the P/S should check the entire schedule to validate that there are no incomplete activities/milestones that are reflected that are prior to the status date.

Some types of schedule maintenance may require certain provisions of the schedule control processes discussed in other sections of this document to be addressed. Those types of maintenance issues will be addressed in more detail in those other sections.

6.3.1 Existing Task Revision

Three types of existing task/milestone revisions may be made. These types are data revisions, informational changes, or task/milestone deletion.

Revising task/milestone data may entail a change of task/milestone dates, duration, resource allocation, or location within the data structure (i.e., WBS). Actual task/milestone dates (start or finish) should be entered as accurately as possible. Planned task/milestone dates should be calculated by the scheduling tool used, not manually entered by the tool user. Duration, or remaining duration, should be updated as required to reflect the most current estimate. Resource allocation changes should also reflect the most current estimate. When changing a task/milestone location within the data structure, ensure that existing interdependencies with other tasks/activities are still valid and accurately reflected.

Informational changes are revisions to task/milestone notes, descriptions, or coding data. If a change control process is in use, it should be consulted and followed for these revisions. If a change control process is not in use, it is recommended that changes of this nature be done in a consistent fashion and in keeping with the existing project guidelines for task nomenclature and coding. It should be noted that schedule change control is considered a "best practice" and should be implemented within each project. The baseline schedule will quickly diverge from the "real" schedule without a change control process. Before deleting any task or milestone from the IMS, existing network logic interdependencies and resource allocations should be reconciled.

6.3.2 Adding New Tasks

New tasks and/or milestones may be added to better define existing work scope or to add new work scope. Both scenarios require adherence to existing schedule controls for descriptions, structure, coding, network logic, duration, resource allocation, and risk data. Oftentimes the addition of new tasks/activities results in a longer duration for the overall schedule, making it necessary to address some of the analysis functions discussed in Chapter 7 of this document.

6.3.3 Logic Modifications

Minor modifications to network logic are necessary on occasion to maintain an accurate reflection of the work being performed. It is highly recommended that a change control process be established and adhered to for logic changes that result in impacts to contractual or other management control milestones. Logic modifications have a direct impact on the planned (calculated) dates for activities, including contractually required and management-directed events. Before and after any significant logic modification, an electronic copy of the schedule should be made and stored for safekeeping. A record of the change should be kept, along with the reason for the change, and the person authorizing the change, particularly if the change impacts a baselined activity or milestone reflecting a program or project commitment. Most automated schedule management tools provide user-defined fields where remarks/comments associated with logic changes can be recorded.

6.4 Schedule Data Back-up and Archive

Program/project schedules should, at a minimum, be backed up monthly or prior to any major changes in the schedule (e.g., logic sequence, task/milestone additions, deletions). This practice ensures that a record of changes is maintained for a number of beneficial reasons. Backup may include electronic and hard copies, but electronic copies at a minimum are recommended. Backup data should be verified periodically and stored in a secure location with controlled access.

Electronically archived schedules can provide a wealth of information such as, work flow history and actual or "as-built" activity durations. Care should be taken to properly label and store these data sets. Conversion of these data sets to upgraded repository software should be maintained to always ensure availability to historical schedule data in later years. The two most important versions of the project schedule would be the original baseline and the "as-executed" schedule. However, electronic schedule versions at key phase changes or events, or major baseline versions, also provide much useful information. As is the case with the lessons learned, archived schedules should be placed within a master database or commonly accessible repository. This provides a source of historical schedule duration information for future duration estimating and validation.

Chapter 7: Schedule Assessment and Analysis

7.1　Overview

Schedule assessment is the process of determining schedule validity and performance at a given point in time. A thorough schedule assessment using many of the techniques described in the following paragraphs should always be performed prior to establishing the IMS baseline. Periodic assessment is also necessary to gain assurance that the IMS continues to generate valid data and to support the project's objectives throughout the project life cycle. A reliable schedule assessment checklist is an important aid that can benefit a project team or outside review team in determining schedule validity. A thorough Schedule Assessment Checklist example is contained in Appendix G.

Schedule analysis is the process of evaluating the magnitude, impact, and significance of actual and forecast variances to the baseline and/or current schedules. After routine updates, schedule analysis begins with the calculation of the critical path and the determination of any change in the completion date of the project. Analysis continues by evaluating schedule performance metrics derived from the IMS and by using this information to assess project health. Analysis results should be reviewed with the project team. This process may be iterated as needed.

IMS assessment and analysis is the same during and after schedule development, with the exception of progress evaluation. The processes that follow should be continued routinely throughout the project life cycle.

7.2　Levels of Insight

The goal for applying the appropriate schedule insight penetration strategy is to enhance the probability of mission success for NASA programs and projects. Mission success must be achieved within the workforce, budget, and time limitations that are levied through all phases of development and operation. Analysis begins with an assessment of the complexity, maturity, and risks of the project being evaluated. Different levels of insight are appropriate when taking these factors into consideration throughout the project life cycle.

Typical project penetration levels as they relate to schedules include the following:

- **Level 0 (No Penetration)** - Accept performing organization's tasks at face value (no additional assessment is required).

- **Level 1 (Low Penetration)** - Participate in reviews and Programmatic Interchange Meetings and assess only the schedule data presented. Perform random spot check assessment and analysis of schedule data.

- **Level 2 (Intermediate Penetration)** - Monthly involvement to spot check, monitor, identify and resolve schedule issues.

- **Level 3 (In-depth Penetration)** - Weekly or Monthly methodical review and analysis of schedule detail.

- **Level 4 (Total Penetration)** - Perform daily or weekly complete and independent evaluation and analysis of schedule detail.

As stated above, many factors affect the level of penetration required in the area of schedule evaluation and analysis. The following list provides a sampling of key factors involved in making this determination:

- Technical risk levels
- Amount of confidence in the performing organization's management abilities
- How well the project planning and control processes are defined and followed
- Project public visibility and impact of failure
- Design complexity, manufacturing complexity, and the ability to be produced
- Value of asset
- Past cost and schedule performance

The project team must consider all the above factors when developing the IMS and when establishing the processes, practices, and guidelines that will be followed in the on-going maintenance, assessment, and analysis of project schedule data. IMS penetration, as it relates to the above factors, is illustrated in Figure 7-1 by mapping the penetration indicator on a standard five-by-five risk cube.

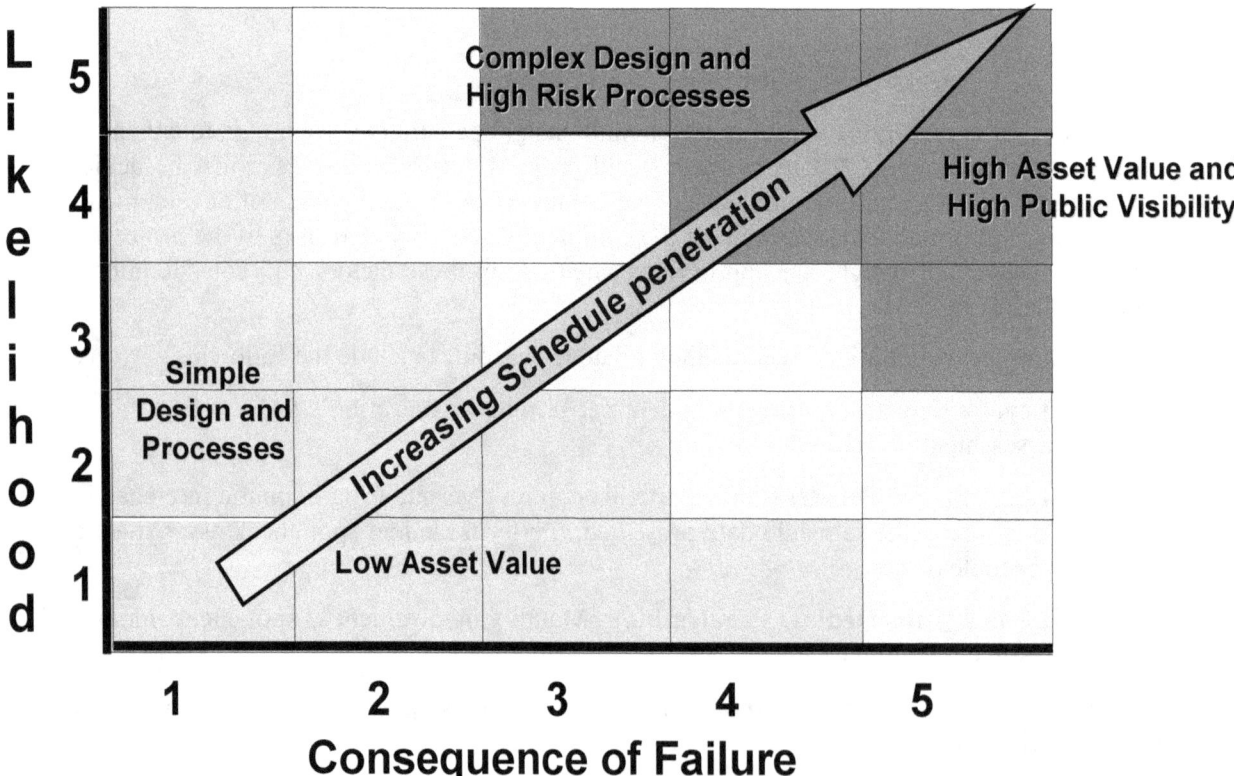

Figure 7-1: Schedule Insight Penetration Mapped on Risk Cube

7.3 Schedule Logic Credibility Health Check

Schedule credibility is determined by monitoring key indicators within the IMS that reflect both good and poor characteristics of schedule structure and maintenance. Examples of key indicators within the logic network that must be monitored include the following: number of missing predecessors and successors, invalid task constraints, omission of task status, improper status on future tasks, logic ties to/from summary tasks, inaccurate logic ties, and improperly reflecting tasks as milestones. These indicators are based on standard rules of logic network development utilized in critical path method (CPM) scheduling techniques.

The indicators noted above should be identified and tabulated routinely from the remaining IMS detailed schedule tasks and milestones (not summary or hammock activities). This is normally accomplished by using the appropriate data-filtering capability provided by the automated management tool being used. Evaluating the number of key indicators will provide quick insight into the quality of the IMS (see Figure 7-2). Critical Path Method (CPM) scheduling guidelines call for logic networks to be structured so that all detailed tasks and milestones have accurate predecessor and successor relationships assigned. Additionally, it is crucial for only valid task date constraints to be used in a logic network, as well as an accurate reflection of current status (including new forecast dates for behind-schedule tasks) for each task and milestone in the IMS database. It is imperative that no uncompleted task or milestone be shown prior to the current status date in the IMS database. The higher the number of instances where these guidelines are not followed in the logic network, the more improbable it is to accurately identify the true critical path in the project schedule. It also indicates that the overall schedule lacks credibility in the data that it produces. This assessment process additionally provides the basic statistics of the IMS content such as current number of total tasks, number of completed tasks, number of remaining tasks, current completion date, status date, and the number of remaining work days in the schedule. This information should be compared after each update to aid in understanding what changes have occurred since the last IMS update.

The figure below (Figure 7-2) illustrates a schedule assessment product that utilizes the process outlined above and applies a stoplight credibility rating based on the number of key indicators tabulated from the IMS logic network. A sample set of stoplight rating criteria is also listed below to provide guidance in determining what stoplight rating should be applied to the schedule. The assessment results should always be explained to the project manager and his appropriate team members to help them get the weaknesses corrected so that the IMS can serve as a credible management tool.

Project Name: Project XYZ		Overall Rating	
Contractor: ACME Engineering		1.4	R
		Current	
Schedule Status			
Current Start (earliest activity Early Start date)		1/1/2005	
Current Finish (latest activity Early Finish date)		3/16/2008	
Approximate Remaining Work Days		722	
Is schedule externally linked to other schedules?		N	
Status Date		6/15/2005	
Task & Milestone Count (excl. Summary Tasks)	Count	% of Total	
Total Tasks & Milestones	192		
Completed Tasks & Milestones	13	7%	
To Go Tasks & Milestones	179	93%	
Logic (excl. Summary & Started/Completed Tasks)			
Tasks & Milestones Without Predecessors	75	42%	R
Tasks & Milestones Without Successors	73	41%	R
Constraints (other than ASAP) and Deadlines	102	57%	R
Summaries with Logic Ties **	1	1%	G
Tasks & Milestones Needing Updates	21	12%	R
Actuals after Status Date	2	1%	Y
Tasks marked as Milestones (have Duration > 0)	0	0%	G
Additional Schedule Information			
Tasks with No Finish Ties	20	11%	
Recurring Tasks	0	0%	
Tasks & Milestones with Estimated Durations	15	8%	
Schedule traceable to WBS (Y/N)	Yes		
Realistic Critical Path(s) (Y/N)	No		
Schedule Baselined (Y/N)	No		
Resource Loaded (Y/N)	No		
Tasks & Milestones with 10 days or less TF	1	1%	
Tasks with Total Float > 25% of Rem Dur	148	83%	

Figure 7-2: Schedule Health Check Example

The following reflects the recommended stoplight criteria used in the above Schedule Health Check rating Process:

Schedule Health Check Rating Criteria

- For missing predecessors, successors less than 5% is green, from 5% to 10% is yellow, and greater than 10% is red.

- For Constraints and Deadlines, less than 10% is green, 10% to 15% is yellow, and greater than 15% is red.
- For tasks needing updates, actual starts/finishes after the status date, and tasks marked as milestones 0% is green, greater than 0% up to 5% is yellow, and over 5% is red.
- For summaries with logic ties less than 2% is green, 2%-3% is yellow, greater than 3% is red.
- The overall project rating is determined by assigning a numeric value to the different colors, i.e., red = 1, yellow = 2 and green = 3.
- The numbers are summed and a weighting factor is applied to determine the final results. The average results are color coded as follows: Red is less than 1.75, Yellow 1.75 to 2.5 and Green greater than 2.5.

Weighting for Overall Schedule Rating

- Missing Predecessors = 20%
- Missing Successors = 20%
- Constraints and Assigned Deadlines = 15%
- Summary tasks with logic ties = 10%
- Tasks and Milestones Needing Status = 20%
- Actual starts/finishes after the Status Date = 10%
- Tasks marked as Milestones (but have Duration > 0) = 5%

7.4 Critical Path Identification and Analysis

The schedule may become very dynamic during the implementation phase, and because of this, it is imperative to always know what sequence of tasks is the real driver affecting project completion. It is also important to monitor the consumption of schedule margin that may exist as part of the critical path. Management insight into the critical path is essential in making accurate resource and manpower decisions to successfully achieve project completion.

Critical path identification and analysis involves constant review of the validity of included tasks, durations, and types of relationships that are involved in the primary critical path, as well as near secondary paths. Often changes made to durations and/or logic relationships can be made to shorten the critical path and prevent project completion from moving to the right.

It is extremely important to note the difference between critical path activities and "critical activities" as defined by management. These two may be, but are not necessarily, the same. In scheduling terms, the critical path is the sequence of activities that are tied together with network logic that have the longest overall duration from time now until project completion. Critical activities may be defined as any tasks which have been deemed important enough to have this distinction assigned to them.

Common characteristics of a credible critical path include the following: it typically begins at "time now" and proceeds to project completion, the tasks and milestones are tied together with network logic in a sequence that is programmatically feasible or otherwise makes sense from a workflow standpoint, the path contains no level-of-effort (LOE) or summary activities, and there are no gaps in time between tasks that cannot be explained. It is recommended that after each update cycle of the IMS, the critical path should be identified and compared to the previous month's critical path. In making this comparison, it is important to clearly understand what has changed, why it has changed, and again validate that the sequence of tasks passes the common sense test.

With sound IMS structure, the project critical path should be identifiable by isolating the sequence of tasks that has the least amount of total slack (float) as calculated using the Critical Path Method (CPM) technique of scheduling. With most scheduling tools, this can be accomplished through the use of filters or development of exception reports by filtering or isolating only those tasks/milestones that have zero slack. Keep in mind that if the task/milestone that represents the final completion of the project has a hard constraint date assigned to it, then there would be a possibility that the critical path could have a positive or negative total slack value instead of zero. Organizing this information by slack value and by date provides a waterfall format that enables management to clearly see what effort is driving the project completion (see Figure 7-3). Please remember that the project critical path can only be identified accurately if all task and milestone interdependencies are satisfactorily incorporated into the IMS to form a complete end-to-end logic network. This type of schedule structure will allow the schedule management tool to accurately calculate slack values for each task and facilitate critical path identification.

Critical path identification and analysis are essential to ensure that management is focusing the necessary resources on the correct tasks to prevent slippage of the project end date. Close monitoring and analysis of the top 3 to 5 paths is also recommended and will ultimately provide management with the necessary insight to better keep the project under control and on track for successful completion.

Activity ID	WBS	Activity Description	Original Dur	Total Float	Schedule
0001	M	Authority to Proceed (ATP)	0	0	26OCT99
0200	M.2.1	Develop Prelim Avionics Rqmts for PDR	65	0	26OCT99 – 01FEB00
0210	M.2.1	Dev Prelim Avionics Design & Analysis for PDR	90	0	24NOV99 – 05APR00
0005	M	Preliminary Design Review (PDR)	20	0	06APR00 – 03MAY00
0230	M.2.1	Compl Avionics Critical Design & Anal for CDR	150	0	04MAY00 – 07DEC00
0430	M.4.1	Compl Propulsion Critical Design & Anal for CDR	150	0	04MAY00 – 07DEC00
0010	M	Critical Design Review (CDR)	20	0	08DEC00 – 08JAN01
0440	M.4.1	Update Final Propulsion Drawings	45	0	09JAN01 – 14MAR01
0450	M.4.1	Baseline & Release Propulsion Dwgs to Mfg	40	0	14FEB01 – 11APR01
0460	M.4.2	Propulsion Component Fabrication	150	0	12APR01 – 14NOV01
0470	M.4.3	Propulsion Components Assy & C/O	80	0	06AUG01 – 29NOV01
0015	M	MUV Component Fabrication Complete	0	0	30NOV01
0480	M.4.3	Propulsion Components Available to I&T	0	0	30NOV01
0600	M.6.1	Integrate Subsystem Components	60	0	30NOV01 – 27FEB02
0610	M.6.2	Subsystem Functional Testing	65	0	31DEC01 – 03APR02
0620	M.6.2	Full System Verification Testing	45	0	04APR02 – 06JUN02
0025	M	MUV System Integration & Test Complete	0	0	07JUN02
0800	M.6.2	Final Pre-Ship Acceptance Review	10	0	07JUN02 – 20JUN02
0810	M.6.2	Prep & Package MUV Hardware for Shipment	5	0	21JUN02 – 27JUN02
0820	M.6.2	Ship MUV	3	0	28JUN02 – 02JUL02
0030	M	MUV Delivery	0	0	02JUL02

Figure 7-3: Critical Path Example

7.5 Schedule Performance Trend Analysis

Analysis on past performance can provide much insight into future expectations. Studies conducted using data from past projects clearly reflect that performance trends rarely improve after projects reach the 15% completion mark. Therefore, it is imperative that project teams establish sound performance analysis practices from the very start of project implementation. The management team needs as much meaningful and credible performance information as possible to help keep the project on track in order to meet planned objectives.

It is recommended that periodic performance and work-off trend analysis be performed on IMS data as depicted in the graphics shown below. Figure 7-4 reflects analysis data that compares performance rates for accomplishing tasks in the past to the quantity of planned tasks required in future months. Caution should be taken if this analysis reflects an unrealistic bow-wave of tasks scheduled to occur with higher required completion rates than the project has been able to accomplish previously. This situation is indicative of a schedule that is most likely unrealistic. In this type of analysis, if the completion rates projected for tasks scheduled for the next six months are much higher than actual completions accomplished during the past six months, then a closer look should be taken at the type of tasks that are scheduled to evaluate the need for re-planning in order to keep the schedule realistic. Figure 7-5 illustrates schedule trend data for cumulative task "finishes" which compare baseline to actual data. This figure also contains trending data that reflects the baseline execution rate (BER). This value indicates a monthly performance rate for accomplishing baseline tasks during the months they were baselined to finish.

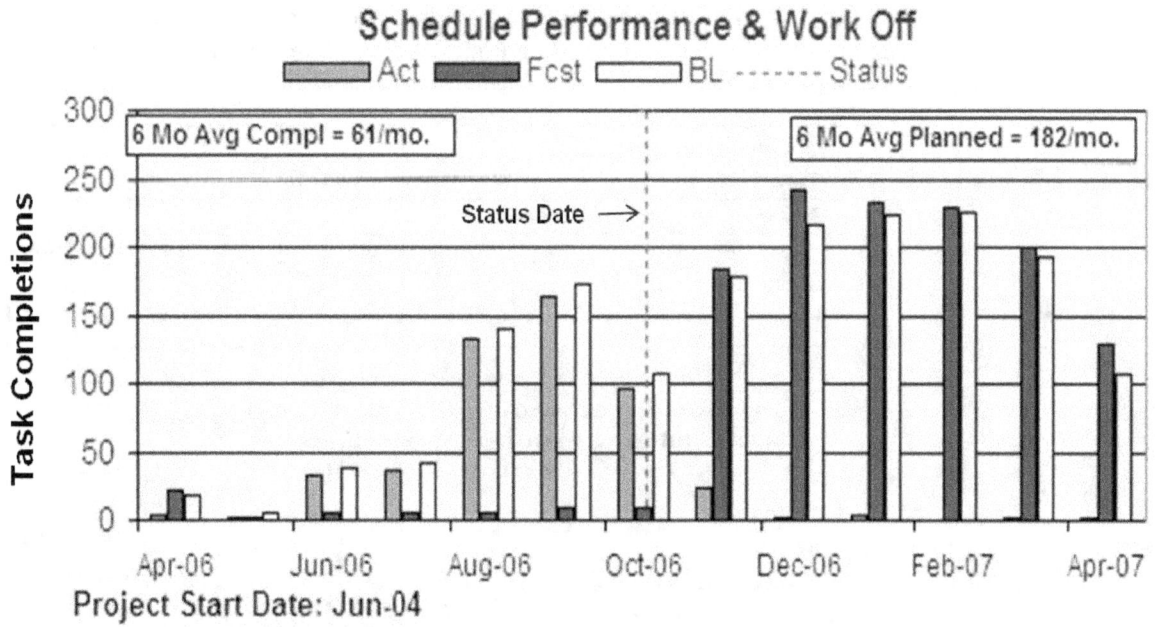

Figure 7-4: Schedule Performance and Work-off Trend Example

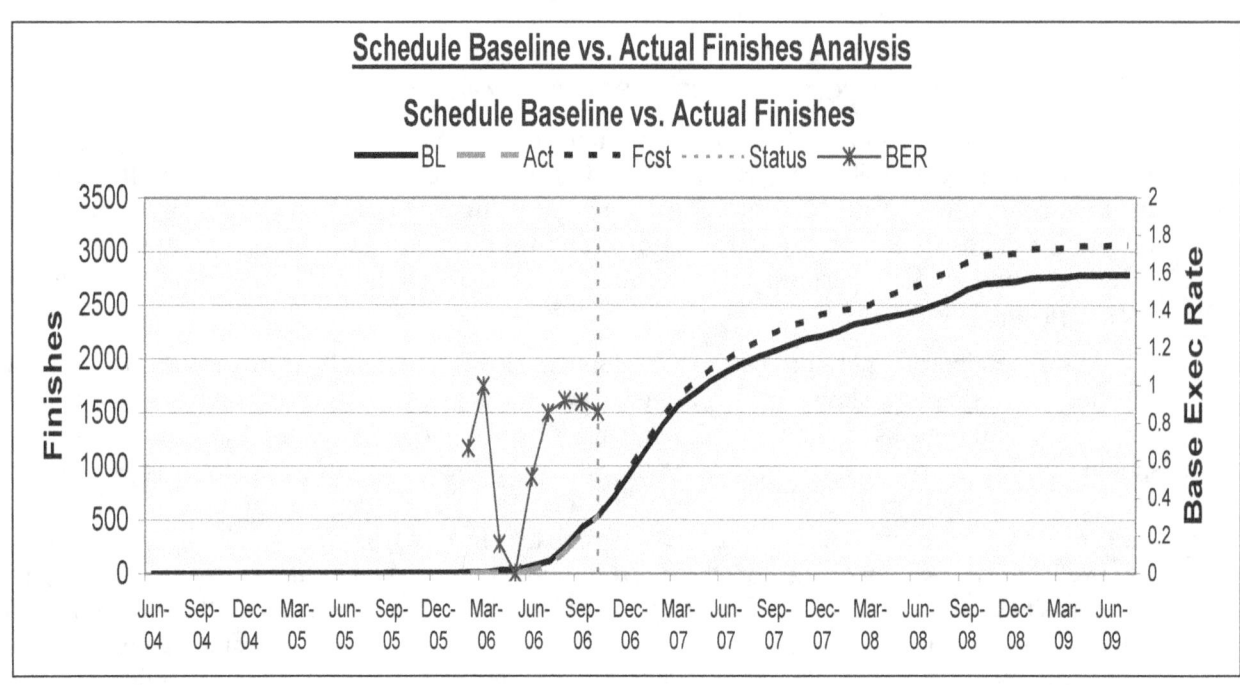

Figure 7-5: Cum Baseline vs. Actual Task Finishes & Baseline Execution Rate (BER)

Another best practice for schedule performance trend analysis uses similar rationale as shown above, but reflects the results as a schedule efficiency factor. The graphic shown below (see Figure 7-6) illustrates how past schedule performance data can be translated into a schedule efficiency rating that can be used to provide insight into schedule health.

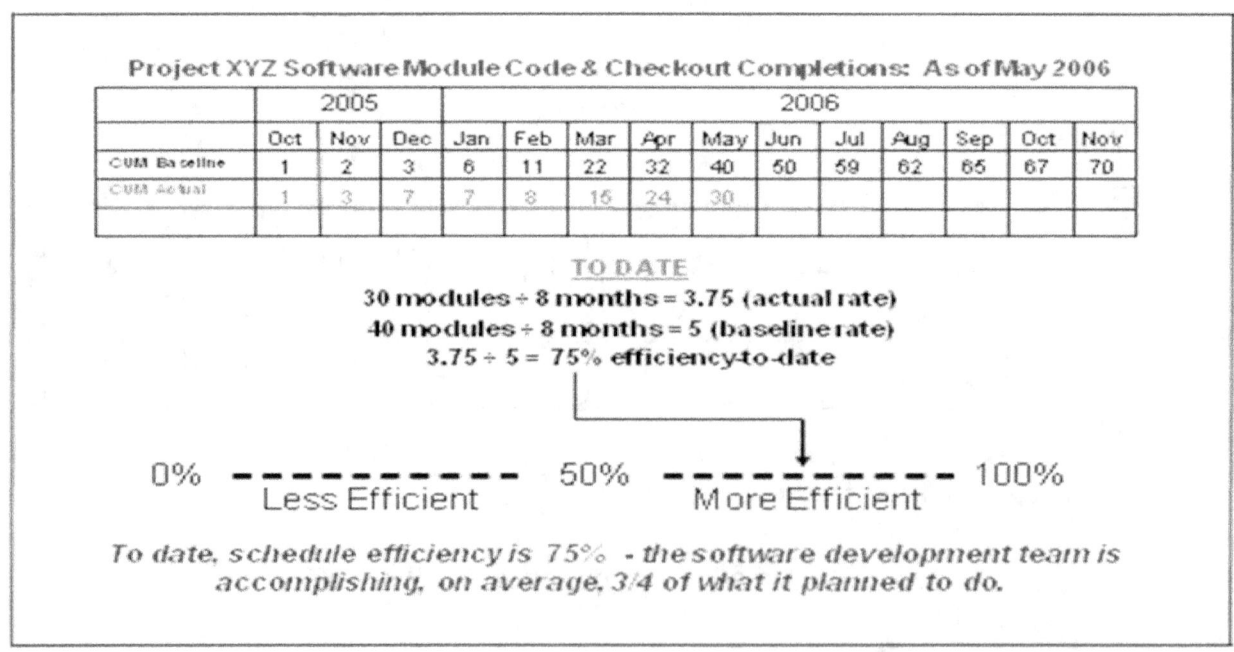

Figure 7-6: Schedule Performance Efficiency Analysis Example

Many other schedule performance trend analysis techniques are available. The graphics that follow (see Figures 7-7 and 7-8) portray two additional recommended formats for assessing and analyzing schedule performance. The rationale and intent reflected in the following figures are very straight forward and can easily be applied to any project IMS.

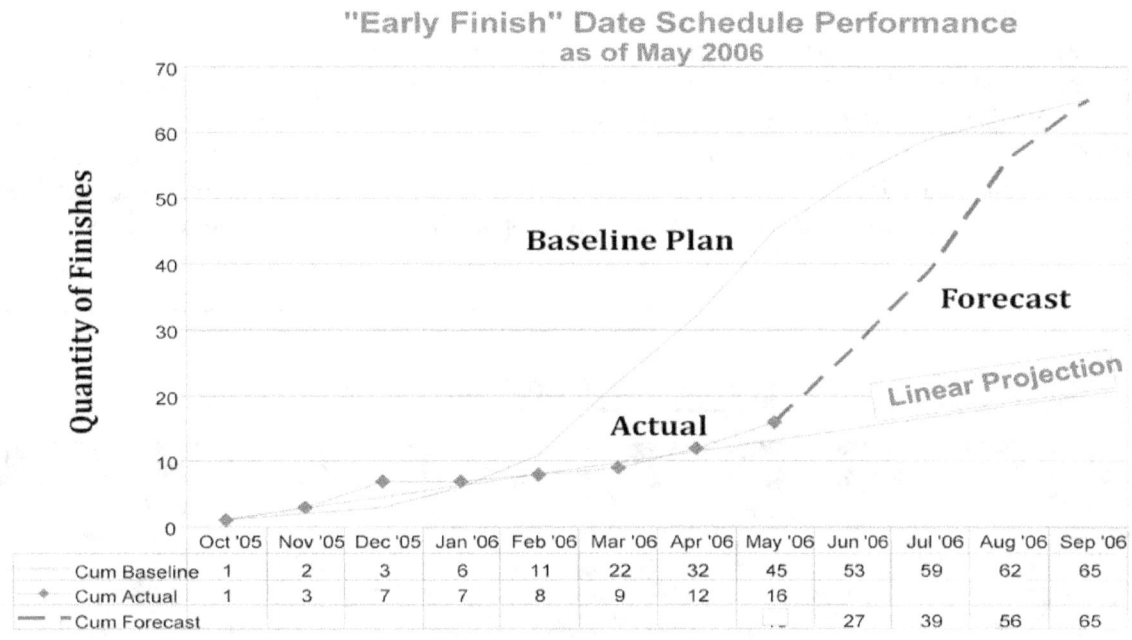

Figure 7-7: Linear Projection of "Actuals" Based on Schedule Performance

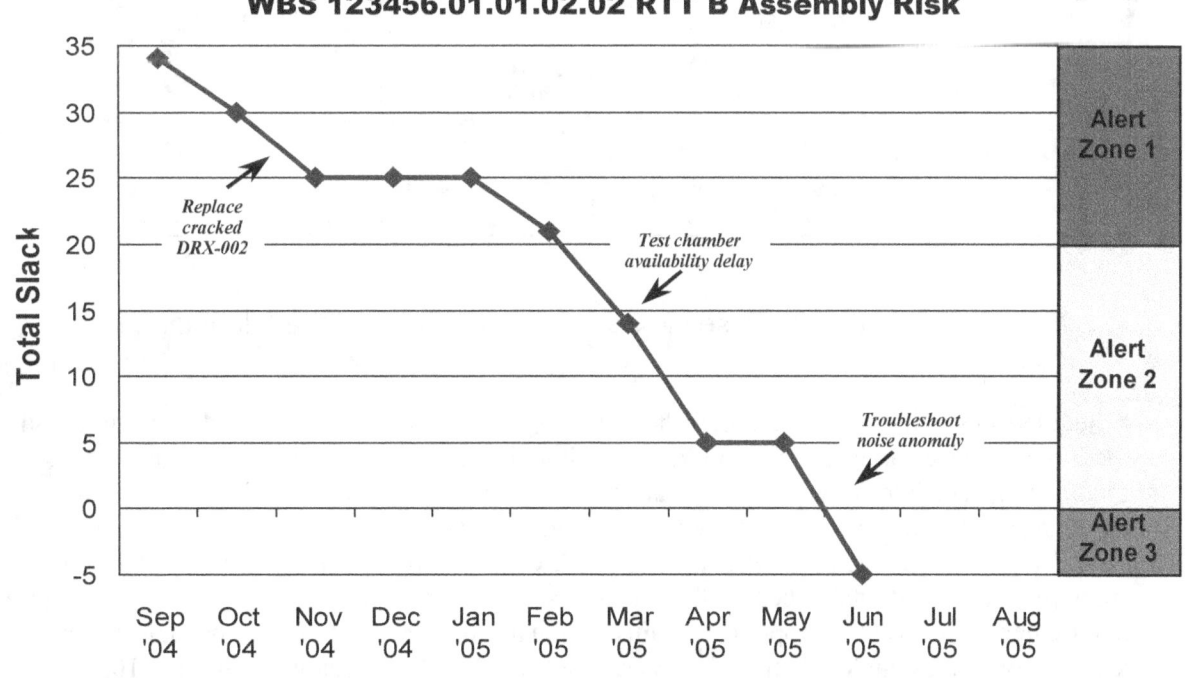

Figure 7-8: Total Slack Trend Based on Schedule Performance Example

7.6 Baseline vs. Current Comparison and Analysis

During the implementation phase of a project, it is likely that the current schedule will at some point deviate from the baseline schedule (see Figure 7-9). This situation is not unusual and occurs for many reasons. It is very important to routinely conduct comparisons between the baseline and current schedule to identify and monitor significant variances, understand why the variances occurred (i.e., the root cause), and what the impacts are to project completion so that appropriate corrective action can be planned. Tasks that have significant deviations from the baseline may also cause a new critical path or a near secondary path. Resources may need adjusting to accommodate these variances and/or work around plans developed to correct the situation. Please keep in mind that the schedule baseline must correspond to the resource baseline and the earned value baseline, so that if drastic deviations occur between current and baseline schedules, it may signal the need for re-planning and/or additional resources.

Schedule Milestone Comparison: (Weighting 25%)

ID	WBS	DESCRIPTION	BASELINE	CURRENT	VARIANCE
26	0	Systems Requirements Review (SRR)	7/11/2006	7/11/2006	0
27	0	Preliminary Design Review (PDR)	9/8/2006	9/8/2006	0
30	0	Critical Design Review (CDR)	2/20/2007	4/9/2007	35
32	0	System Test Readiness Review (STRR)	10/15/2007	10/15/2007	0
33	0	Space Vehicle I&T Start	9/18/2007	11/19/2007	45
36	0	Space Vehicle I&T Complete (Sell off com	8/28/2008	8/28/2008	0
38	0	Flight Readiness Review (FRR)	10/21/2008	10/22/2008	2
39	0	Launch Readiness Review (LRR)	10/27/2008	10/28/2008	2
40	0	Launch	10/28/2008	10/28/2008	0

Figure 7-9: Baseline Schedule vs. Current Schedule Example

Schedule thresholds should be established by the project management team that aid in identifying and focusing on those variances that should be monitored and managed. Two key factors to be considered in establishing schedule variance thresholds are the number of days a task/milestone has changed from the baseline schedule and also how many days of total slack are associated with those variances. Thresholds may also vary due to the type, length, and complexity of the project being implemented. Thresholds that are agreed upon and established by the project management team should also be applied with the added requirement to provide appropriate variance rationale from the responsible manager or implementation team. A typical example of a Schedule Variance Report is shown below in Figure 7-10.

SCHEDULE VARIANCE REPORT

Date: 6/15/04
Project: Research Rack #1
WBS #: 123456.01.01.02   Avionics
           123456.01.01.02.01   Master Controller
Schedule ID#: 536
Description: Power Control Assembly Complete
Baseline: 7/30/04
Forecast: 9/15/04
Cause: Cable connectors delivery date slipped 1 month by vendor
Corrective Action: Locate and install temporary connectors to allow for powering-up of unit for testing as previously scheduled

Figure 7-10: Schedule Variance Report Example

7.7 Schedule Margin Assessment

Adequate schedule margin appropriately placed in a project schedule is critical to project success. A probabilistic schedule risk assessment is highly recommended as a basis for determining adequate schedule margin. Schedule margin should be easily identifiable and strategically placed within the IMS. Generally, it is recommended to create specially labeled tasks for schedule margin and place the bulk of margin at the end of the schedule just prior to project completion so that it will be reflected and easily accounted for and managed as part of the critical path sequence. Other smaller blocks of schedule margin could also be associated with significant key events in the IMS and placed logically just prior to those events. Please note that when smaller blocks of margin are created and associated with key events within the IMS, they may not fall on the project critical path and therefore will have no effect on the project completion date.

Throughout the project life cycle, it is important to monitor schedule margin. It is good practice to maintain a log indicating the changes in schedule margin and the reason for those changes (see Figure 7-11). Monitoring this metric and comparing it to critical path float not only gives an indication of schedule progress at a high level, but also provides an indication of how optimistic and/or realistic the schedule completion date is. Margin is a key factor to be considered when performing schedule risk assessments.

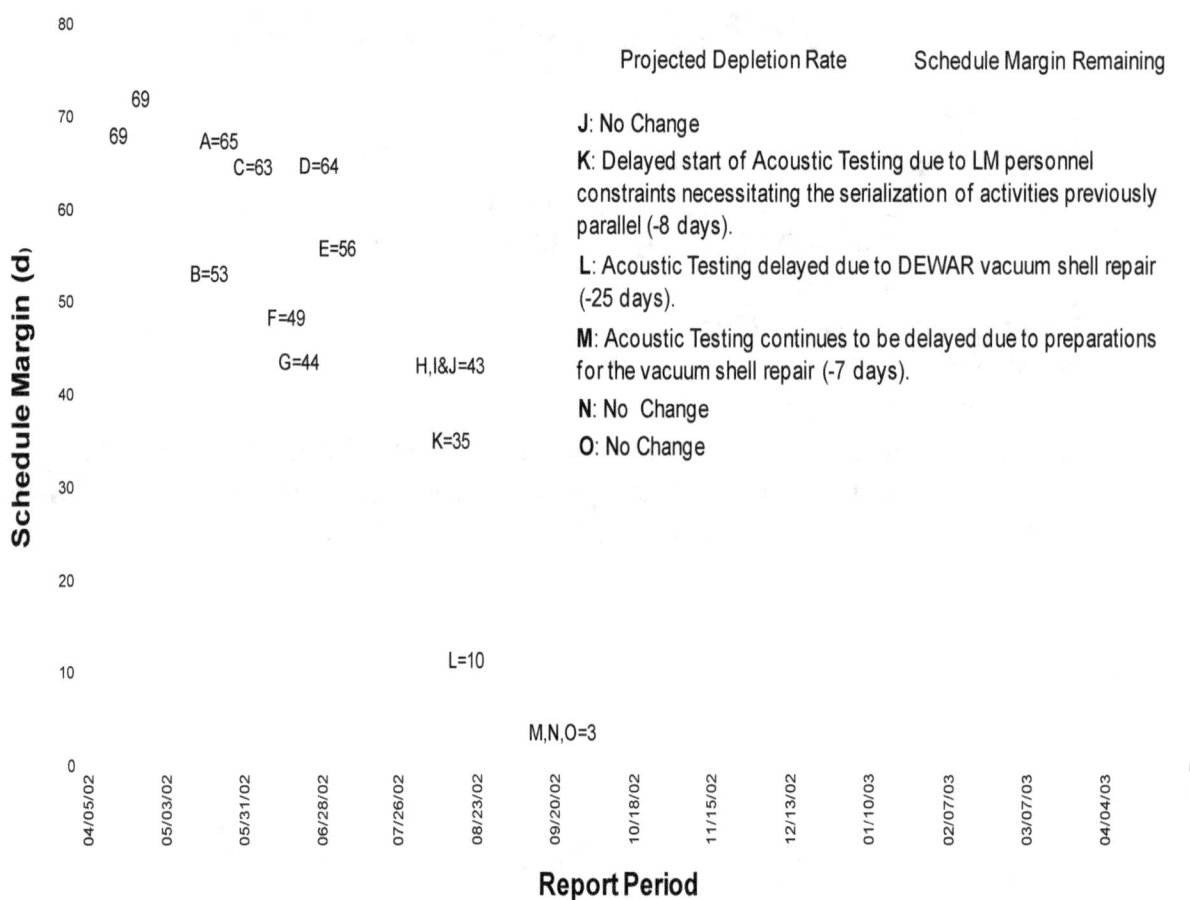

Figure 7-11: Schedule Margin Log Example

7.8 Validate Cost/Schedule Integration

NPR 7120.5 requires that the project's schedule baseline be integrated with the budget and technical baselines to form an overall project integrated baseline. This correlation is essential to ensure that adequate resources are available to accomplish the work when it is scheduled. Without this correlation and validation the project IMS loses credibility.

While resource loaded schedules are not an Agency requirement, they are strongly recommended to ensure that cost/schedule correlation exists. This process also provides a means for in-depth, integrated cost/schedule analysis, at all levels of detail. Analysis using this type of integrated data can significantly aid management in determining whether or not the resources assigned are sufficient to complete the project on schedule and within budget and also prevent resource conflicts. Resource loading of the IMS can only be accomplished through the use of an automated schedule management tool that provides functionality for CPM schedule management and for the assignment of resources at individual task levels. When this process is properly executed, the capability for resource leveling and summarization will also be available to provide added assistance in the validation of cost/schedule correlation and sufficiency. Figures 7-12 and 7-13 shown below, illustrate simple examples of detailed resource loading and leveling and the benefits this process brings to the management team in analyzing resource adequacy and assessing impacts to schedule forecasts. Since resource and workforce planning are heavily dependent on the IMS for time phasing, it becomes critically important for technical managers, financial analysts, and schedulers to maintain constant coordination with regards to cost, schedule, and scope integration.

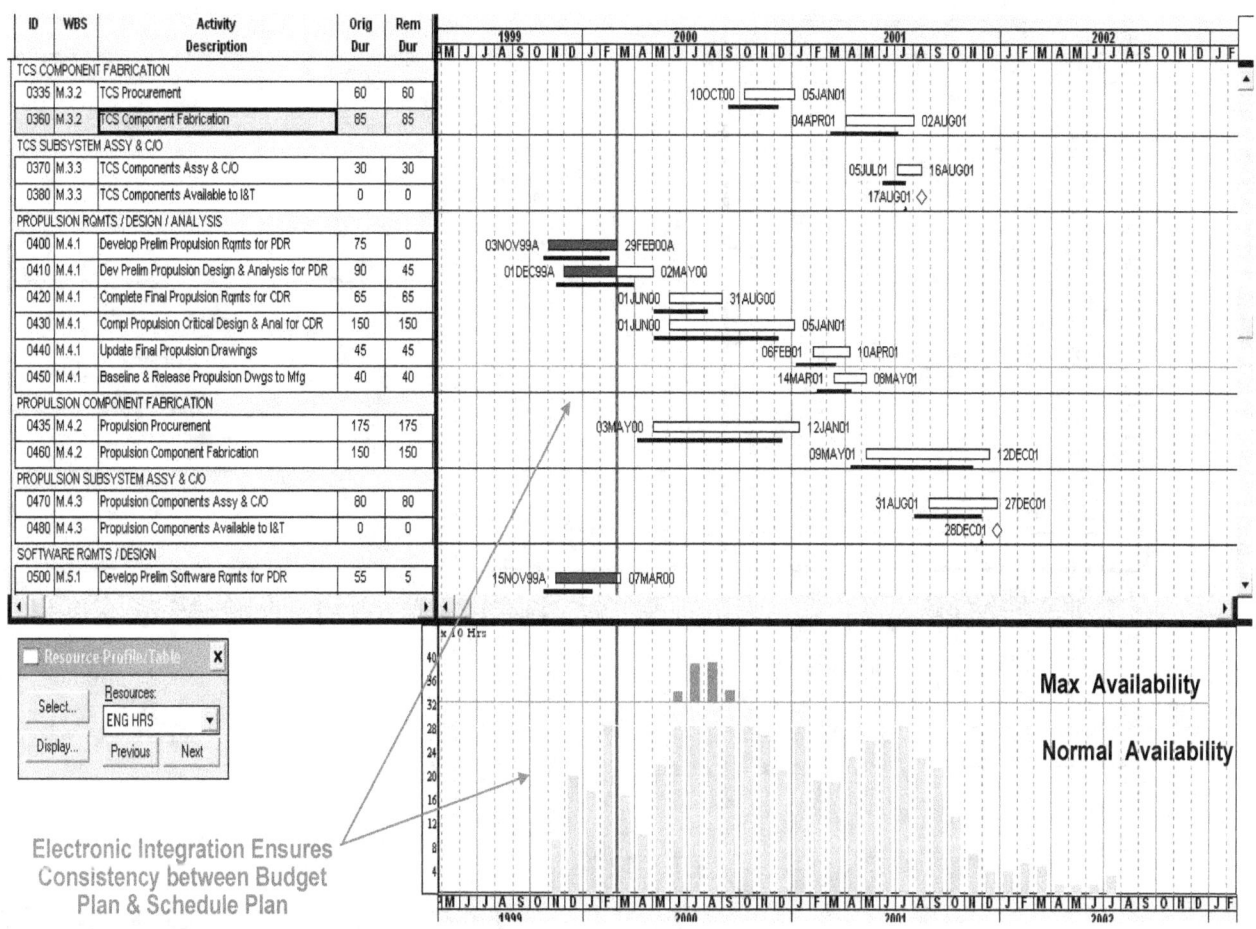

Not resource leveled

Figure 7-12: Resource Loaded IMS with Resource Conflicts Example

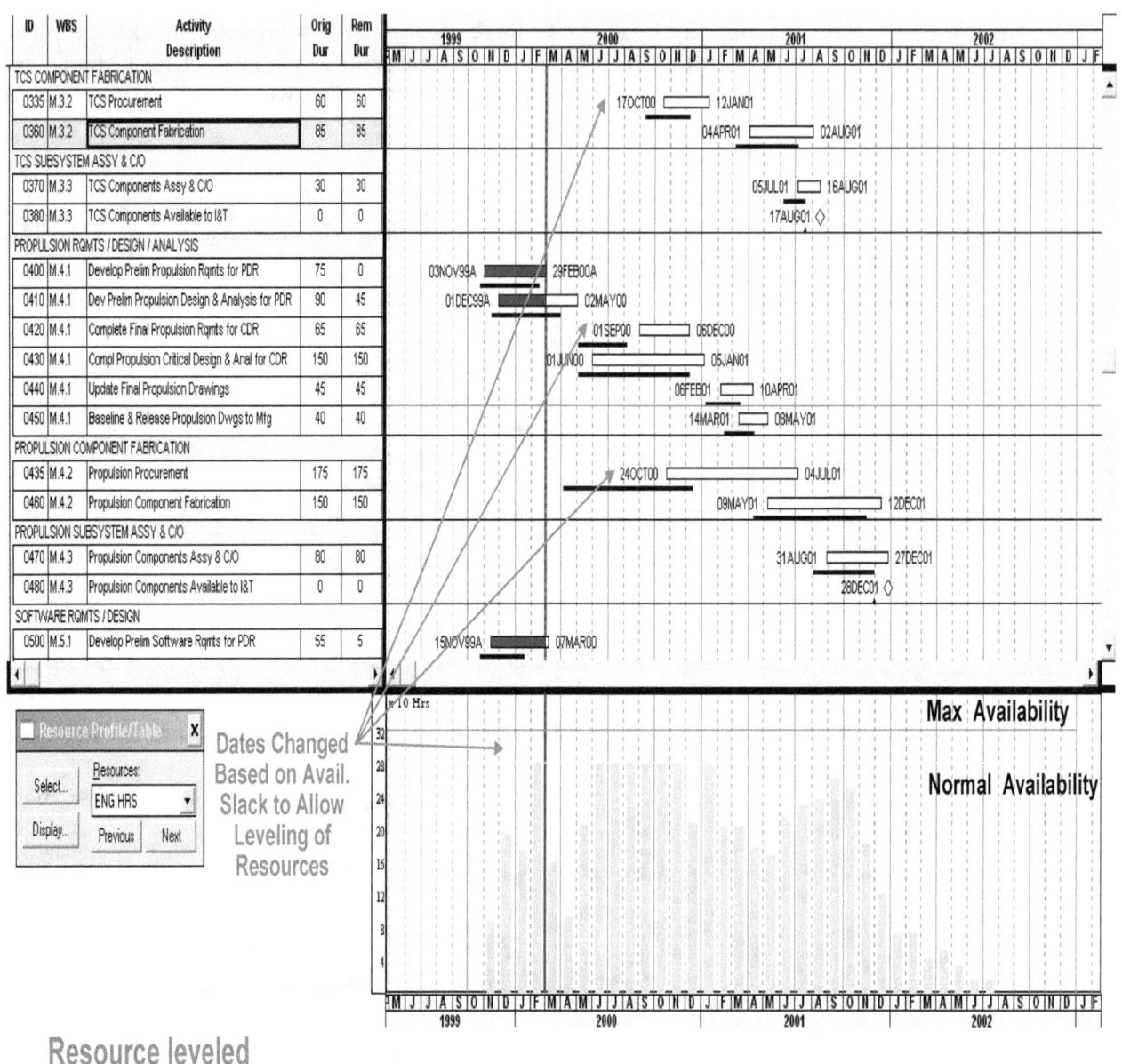

Figure 7-13: Resource Loaded IMS with Leveling to Resolve Conflicts

Another recommended analysis technique is a summary level cost/schedule correlation check (see Figure 7-14). From a summary level perspective, this technique will provide insight that will aid the management team in determining if the resource plan is consistent with the schedule plan. This process entails overlaying the total cumulative project level budget/resource plan on top of the project timeline with only the major milestones indicated. This graphic will clearly reflect budget/resource peeks in relationship to the key milestones and help management determine if the peeks are occurring at the appropriate time in the project plan. For a typical development project, the resource peak should occur a short time prior to the Critical Design Review (CDR). This is due to the required overlap of many skill types during the transition from hardware design to hardware fabrication and test. If this analysis indicates the peek is too early or late, the management team should review and adjust the budget/resource plan accordingly.

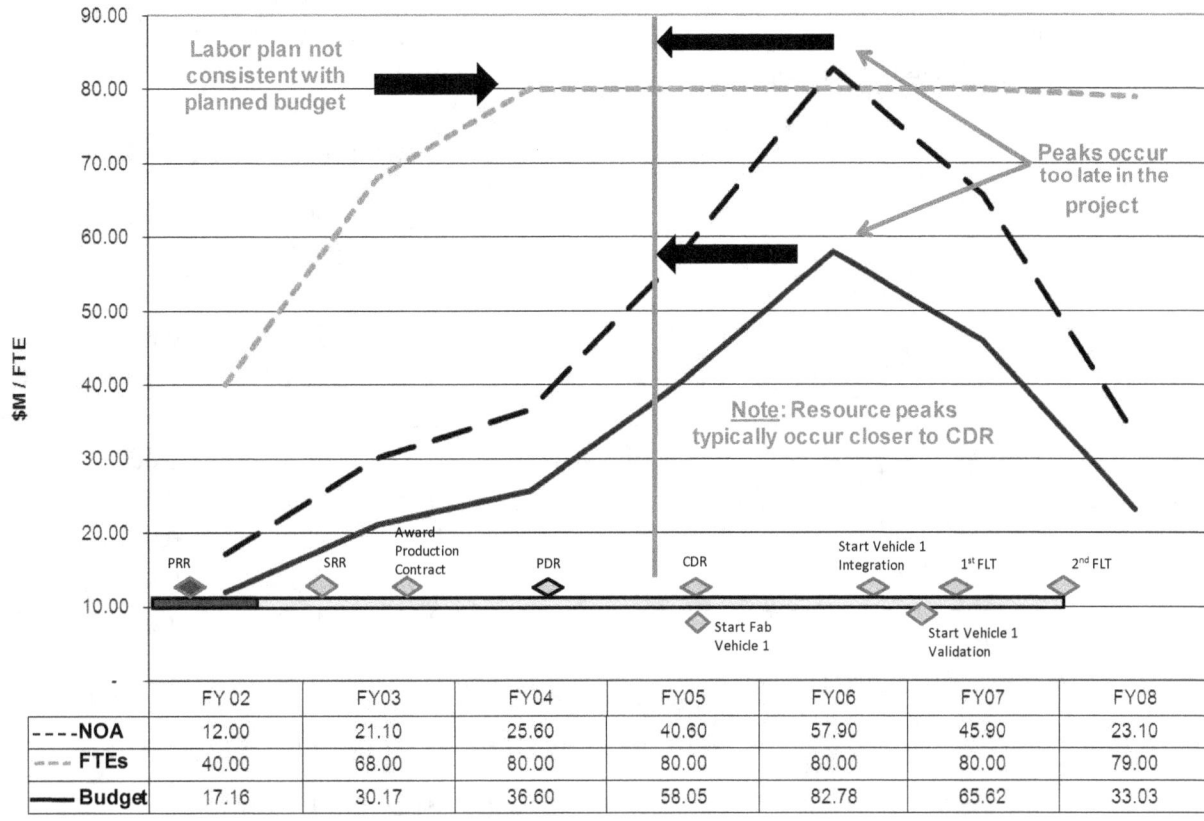

Figure 7-14: Summary Level Cost / Schedule Correlation Check Example

7.9 Schedule Risk Assessment (SRA)

Conducting a SRA is crucial during project formulation and throughout the on-going implementation life-cycle of a program/project. Although there are various ways of evaluating cost and schedule risks and confidence levels, a recommended technique is through the use of a proven probabilistic risk assessment tool with random sampling functionality. The SRA is an important analysis process that evaluates the likelihood that a project plan, reflected in the IMS, is achievable within the planned finish date constraints. This "best practice" approach utilizes an IMS with all work scope reflected in discrete tasks/milestones with all task interdependencies included. The random sampling functionality computes project schedule data simulations based upon realistic duration estimates and a probability distribution parameter assigned to tasks within the schedule. The duration estimates that are associated with schedule tasks include the minimum, maximum, and the most likely durations expected for each schedule task. The probability distribution parameter, also known as the Probability Distribution Curve (PDC), models the likelihood of task durations being successfully achieved between the minimum and maximum parameters. As numerous simulations are executed and calculated, the SRA tool will factor into the random sampling calculations the assigned duration estimates, the selected distribution probability curves, the assigned task interdependencies, and known valid constraints to provide probability percentages for completion dates of all or selected tasks within the IMS (see Figure 7-15).

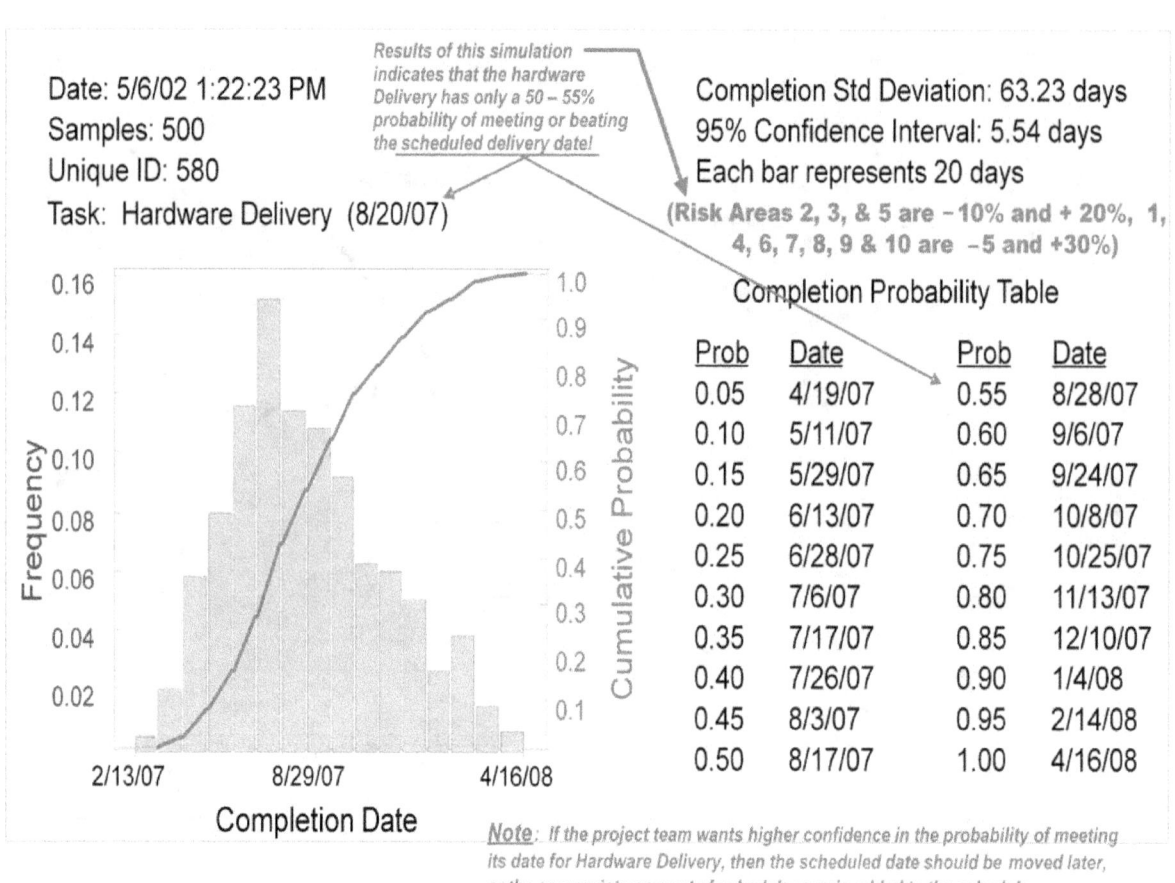

Figure 7-15: Schedule Risk Assessment Example

The process described above will assist the management team in assessing both the credibility and confidence level of schedule data contained in the IMS. The information generated by the probabilistic SRA determination also aids the management team in determining an adequate amount of cost reserves and schedule margin to be planned for prior to establishing an integrated project baseline. Another assessment result of the SRA is the identification of risk critical paths in the project. These paths may or may not be the primary critical path within the project, but due to their associated risk they could, at some point, become the driver to project completion. The identification of these paths will allow the project management team to proactively manage this work to minimize the impact to the project.

It is extremely important for project P/Ss to collaborate with responsible technical team members to gain as much risk and uncertainty related information as possible for their tasks to assist in establishing realistic parameters used in the SRA. During this process, projects must consider the likelihood and consequences of the discrete risks already documented in their formal risk management tracking system or risk register. By relating the identified project risks to the appropriate tasks within the IMS, more realism can be included in assigning task confidence parameters. This collaboration will lead to a more accurate assessment of cost and schedule risk, both before the IMS baseline is established, and afterward at various assessment intervals.

There are additional assessment techniques available to assist the analyst in creating a more realistic model of the planned effort when special conditions dictate the need. These techniques are typically made possible through use of the same type of probabilistic assessment tool discussed above. The additional techniques are briefly described below:

- *Probabilistic Branching* - addresses situations where there is more than one path forward and the path forward is dependent upon the outcome of a particular task. In the event that the outcome of an accomplished task will result in only one of the successors being executed (with the other successor or successors being alternatives dependent upon the predecessor outcome), the analyst assigns a probability to each "branch" that reflects the likelihood of that branch being the one that will be executed. For example, if the outcome of a predecessor task were equally likely to result in one of two possible successor alternatives, each successor branch would receive a 50% weighting in the probabilistic branching model. One example of a practical application of this technique is with a testing task. A test may be successful or it may fail. Failure would result in a different path forward than a successful test would.

- *Conditional processing* - may be required for situations when a task duration or outcome may be affected by external situations. Typically, the duration of a task is dependent upon two factors: the resources available and the amount of work to be accomplished. In certain circumstances, it is possible that task duration may be affected by other factors, such as the time of year the task is performed. For example, if a task named "Testing" is required to be performed when ambient temperatures are above 50 degrees Fahrenheit, a delay in a predecessor could cause the Testing task to be delayed until colder weather. In this situation, you may want to define conditional processing that specifies an increase in the most likely or maximum duration (or both) in the event the predecessor finishes after a certain date. The means of inserting conditional processing varies based on the software tool used but typically consists of constructing a logical "If...then..." statement attached to one or more tasks.

Continued use of the SRA process is important because of new project risks or uncertainties which may have been identified or changed, in turn changing the ultimate cost and schedule confidence-level probability percentages. Many times, as better information becomes available, or external delays occur, these assessments will also change. Recommended intervals for conducting a SRA include: prior to baselining, at major milestone events, or phase changes, significant schedule changes are being considered, identified risk elements or parameters change significantly, annually, or as often as needed.

7.10 Duration Compression

Frequently, during initial IMS development or in later implementation phases, management may determine that the overall critical path duration must be shortened. When faced with this situation there are two approaches that are typically used. A technique called "Crashing" the schedule may be employed which involves increasing resources and reducing durations on those critical path tasks where the most cost-effective time acceleration is achieved. This approach will definitely result in higher costs so that a time/cost tradeoff should be evaluated when using this method. Another approach called "Fast tracking" the schedule typically involves the identification of tasks on the critical path that can be partially overlapped or possibly even a total parallel implementation. This technique may not always result in higher costs, but definitely could increase risk to the project especially in the area of delays due to rework. Both methods should be considered carefully before using them to shorten the schedule.

7.11 Earned Value Schedule Analysis

Project performance data gained from an earned value management (EVM) system may provide additional management insight into schedule performance. Remember that the EVM process works by integrating the project's budget plan with the approved schedule plan to provide an overall project performance measurement baseline (PMB). As actual costs are accounted for, and actual schedule status is taken, an evaluation can be made whether the amount of dollars spent accomplished the amount of schedule planned. This calculation reflects the earned value performance (see Figure 7-16).

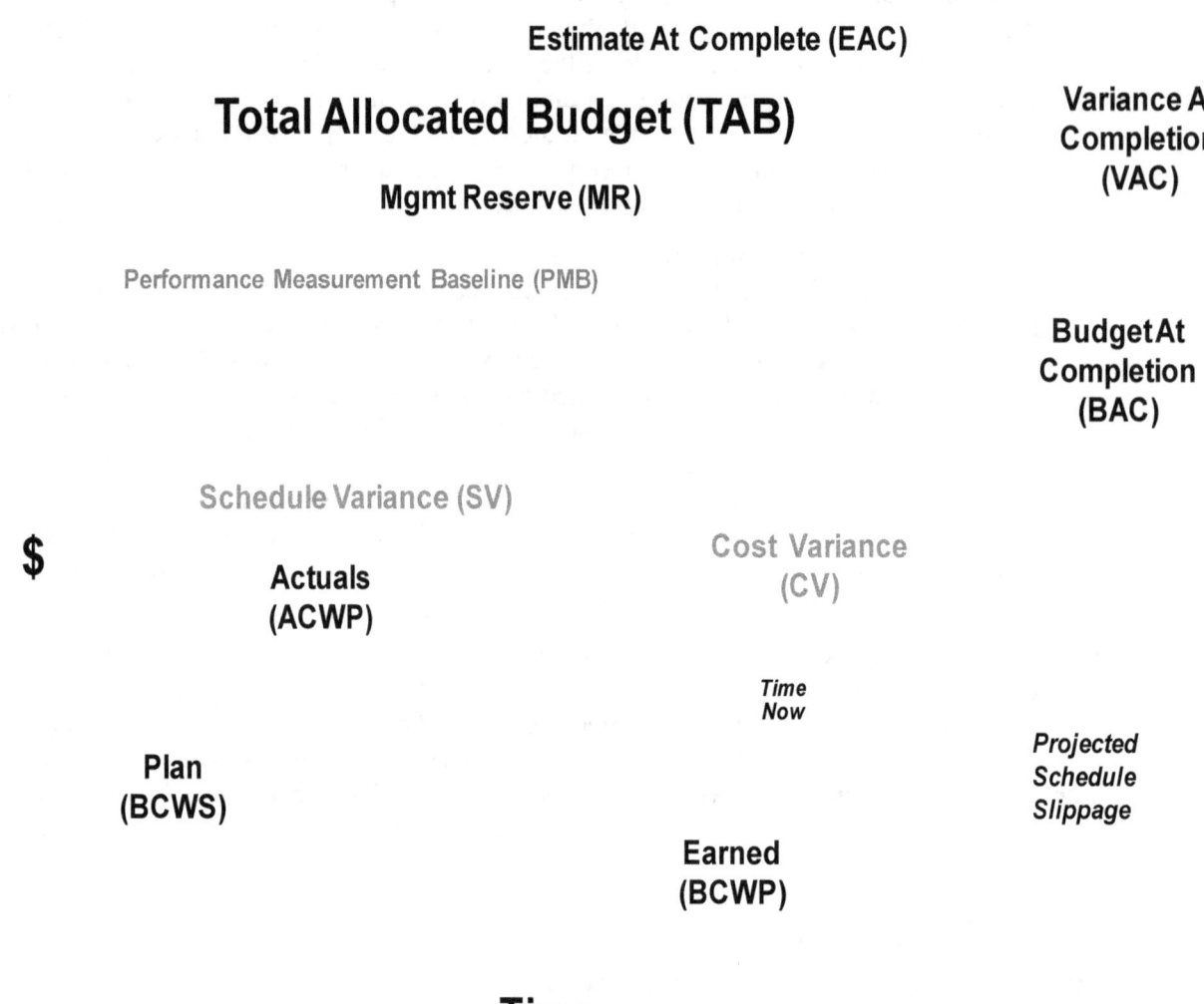

Figure 7-16: Schedule Performance Insight Using EVM Metrics

Schedule Variance (SV) is defined by the following calculation:

Budgeted Cost of Work Performed (BCWP) – Budgeted Cost of Work Scheduled (BCWS)

As the figure above indicates, this can also be stated as the value of the work completed minus the value of the work that was planned to be completed. When this calculation results in a negative answer it means that the project has not accomplished its planned work.

Another important schedule metric that can be calculated from EVM elements is the *Schedule Performance Index (SPI)*. The formula for SPI is the following:

Budgeted Cost of Work Performed (BCWP) / Budgeted Cost of Work Scheduled (BCWS)

This formula provides a ratio of the value of work actually accomplished to the value of work that was planned. When the result of this calculation equals one, then the project is actually accomplishing all work that it had planned. When the result is less than one, then there is less work being accomplished than was planned. And when the result is greater than one, then more work is being accomplished than was planned in the baseline.

The EVM process effectively integrates cost and schedule for each WBS element so that a *Cost Performance Index (CPI)* can also be calculated to provide another means of gaining general insight into project performance. The CPI is calculated using the following formula:

Budgeted Cost of Work Performed (BCWP) / Actual Cost of Work Performed (ACWP)

The above formula provides a ratio of the value of work that was actually accomplished compared to the value of what was actually spent to accomplish the work. When the result of this calculation equals one, then the project is actually accomplishing a dollar's worth of work for every dollar spent. When the result is less than one, this indicates that less than a dollar's worth of work is being accomplished for every dollar spent. And conversely, when the result is greater than one, then more than a dollar's worth of work is being accomplished for every dollar spent.

EVM metrics enable the project management team to easily identify and prioritize those elements of the project that are not performing to their baseline plan. Figure 7-17 reflects how schedule performance can be monitored and analyzed to help the project stay on track.

	WBS	Description	SV	CV	VAC	SPI	CPI	SV	CV	BAC	EAC	% SPENT
1	3200	COMMUNICATIONS	↑	↓	↔	0.777	0.844	-203.2	-130.8	2,043.0	2,130.0	41.03
2	3700	DATA DISPLAY	↑	↔	↔	0.585	1.000	-113.0	0.0	388.0	388.0	41.13
3	3300	AUX EQUIP	↓	↓	↓	0.877	1.133	-93.2	78.2	2,418.2	2,409.8	24.33
4	3100	SENSORS	↑	↓	↔	0.908	0.971	-36.6	-10.6	1,728.4	1,750.0	21.49
5	2100	PROJ MANAGEMENT	↑	↔	↔	0.959	0.942	-12.0	-17.4	618.4	621.6	48.51

(Top 5 Project Schedule Variances per EVM Analysis — columns SV and CV)

Figure 7-17: Schedule Analysis Using EVM Indicators

Combining the CPM analysis of schedule total float (TF) values and schedule performance index (SPI) values provides the analyst with a comprehensive performance view of the status of a project from an integrated cost and schedule prospective. The following table (Figure 7-18) provides a brief insight into how TF values and EVM indices can be effectively analyzed for common project scenarios.

SPI	TF	Scenario
>1	>0	Ahead of schedule
<1	<0	Behind schedule
>1	<0	Critical tasks behind, total work ahead (priority issue)
<1	>0	Critical tasks ahead, but total work behind (priority issue)

Figure 7-18: Schedule Analysis Utilizing Total Float and SPI

While it is true that earned value analysis is another valuable management tool to help provide schedule insight into project health, it should always be used in conjunction with standard CPM schedule analysis techniques, which give the added perspectives of schedule slack and critical path identification. Earned Value trends enable management teams to forecast total project costs and schedule, and routinely monitor deviations from the baseline. This capability can ultimately help management achieve project completion on schedule and within budget.

Chapter 8: Schedule Control

8.1 Overview

Schedule control is a disciplined process for managing baseline schedule data, as well as current IMS content. It should be understood that baseline schedule data represents firm commitments while current schedule data reflects actual and forecasted outcomes. Baseline and current schedule changes are to be clearly identified, evaluated, documented, and approved. Disciplined schedule control provides assurance that the baseline and current IMS includes all of the work authorized and is maintained in a manner that is accurate, traceable, and defendable. Schedule control techniques put in place depend upon programmatic requirements, project size, and complexity of the schedule. Examples of common schedule controls include Baseline Change Requests (BCR), Schedule Change Boards (SCB), Configuration Change Boards (CCB), IMS change logs, and version controls. Each of these will be explained in the appropriate sections that follow. Application of the principles in this section applies equally to all contractor and NASA schedules. However, approvals and notifications may vary depending upon the contract or organization. The project team should refer to the authorizing documents for specific schedule control requirements.

The following definitions and guidance are provided to ensure conformity with current Agency policy for baseline management. The following definitions and guidance apply to NASA and contractor schedules.

Definitions:

- A *Commitment Baseline* establishes and documents an integrated set of project requirements, cost, schedule, technical content, and an agreed-to Joint Cost and Schedule Confidence Level (JCL) that forms the basis for NASA's commitment to OMB and Congress.

- A *Management Baseline* is an integrated set of requirements, cost, schedule, technical content, and associated JCL that forms the foundation for program/project execution and performance measurement.

Guidance:

- Only one official NASA baseline exists for a project, and it is the Commitment Baseline. The Management Baseline is a subset of the Commitment Baseline.

- The Commitment and Management Baselines are mutually set at the appropriate KDP gate as directed by NPRs 7120.5, 7120.7, and 7120.8 and as documented in the program/project Plan,

- Changes to the Commitment Baseline occur via the re-baselining process and require coordination with OMB and Congress,

- Projects go through a re-baselining process which generally includes the IMS when: (1) the development cost portion of the Commitment Baseline is exceeded by 30 percent or (2) when the decision authority judges that events external to the Agency make a re-baseline appropriate, or (3) the decision authority approves project scope changes that can be accommodated within the project's Commitment Baseline, but at the same time have significant impact on the Management Baseline.

- The program/project manager has the authority to re-plan within the approved Management Baseline, but must obtain the approval of the proper decision authority.

8.2 Baseline Content

The first step in establishing schedule control is to determine what schedule content will be formally controlled. In making this determination, consider the external schedule commitments made and where they are included in the baseline. Another important consideration relative to baseline content is whether the project has a requirement to implement EVM processes. For projects not employing EVM, there is usually more flexibility in determining the baseline schedule content. For projects requiring EVM processes there is a more direct and rigorous relationship to be maintained between the total baseline schedule and performance measurement processes.

For non-EVM projects, it is recommended that the project focus the management baseline schedule control process on a carefully selected and meaningful set of tasks and/or milestones from the IMS. This set of control tasks and milestones may include, but is not limited to, items such as contract milestones, major integration milestones, key procurement milestones, critical test activities, hardware deliveries, facility readiness milestones, technical reviews, verification milestones, and operational readiness milestones. Using this approach, the control of the baseline IMS is limited to control of these approved tasks and milestones. Changes to other task and milestone dates will not require approval through the baseline change control process unless their movement impacts a task or milestone that is baseline controlled.

For projects requiring the implementation of EVM processes, the management baseline schedule becomes an integral component for development and time phasing of the performance measurement baseline. The PMB is a time-phased budget based on the cost estimates for resources allocated to IMS tasks. The PMB should remain stable and only be modified due to changes in authorized work scope, authorized re-plans, or authorized re-baseline activity. Therefore, a requested change to future plans contained in the baseline that result in modifications to the task sequence and/or budget phasing for remaining work will require an approved change to the PMB. If these changes are not controlled, documented, and reflected accurately in the revised baseline, then project performance will not be measured accurately and variance identification will be compromised. The impacts could potentially result in additional and unnecessary variance reporting requirements and more importantly, faulty schedule information for management decision-making use.

8.3 Baseline Control Process

Many organizations use a configuration control process to manage various types of technical and programmatic baseline changes. What follows is a description of the typical process sequence for a project engaged in baseline schedule control. This process is used regardless of whether the proposed change is to the external or internal schedule baseline. Terms and details reflected below may vary to some degree between projects. The following is meant to be an illustration of the application of baseline schedule control principles.

(a) A Baseline Change Request (BCR), *or equivalent*, (Figure 8-1) for the IMS is initiated by a responsible program/project Technical Lead, responsible contractor, or other outside customer source.

(b) The responsible change initiator who originates the BCR should coordinate with the project Schedule Office to determine the resulting impacts caused by the proposed change. Impact analysis should be conducted on both external and internal schedule baselines. This may require the preparation of "what-if" versions to assess the impact of the proposed change utilizing either the original baseline plan, the currently approved baseline plan, or the current updated IMS.

(c) The BCR not only documents a clear description of the proposed change, but also the "before" and "after" effects of the proposed change on the internal and external schedule and budget baselines.

(d) The BCR is then brought to the governing project Change Control Board (CCB) and reviewed in accordance with the project's Configuration Management Plan. Note, if the proposed schedule change impacts the external schedule baseline, then review and approval will also be required by the appropriate governing change board at the program, mission directorate, or Agency level,

(e) Once the schedule BCR has been formally approved by all the applicable change control boards, the project P/S issues an updated schedule with a new revision designator assigned to the baseline IMS.

(f) The project P/S should maintain a log of all revisions to the baseline IMS. This log provides the on-going baseline schedule traceability required for sound project configuration control.

Baseline Change Request

Date Initiated: December 1, 2009	BCR# 001
Originator (CAM): John Roberts	WBS Element Number: 123456.08.05.09.01.03
WBS Element Name: Develop prototype software code	
Description (Title) of Change: New software functional requirements added to scope.	
ECR Ref # (if Applicable): 123456-011-99	
BAC Increase/Decrease ($K): $78K	New Period of Performance: 2/3/2010 – 4/15/2010
Scope Change: _X_ Yes __No / Description of Change: Implement new software data extraction, summarization, and repository functional requirements to the software prototype development. Details: Add 6 week task and $78K for labor to do code development, unit test, and debug activities. Apply to Control Account 123456.08.05.09.01.03 / WP 03	
EAC Increase/Decrease ($K): $78K	New Period of Performance: 2/3/2010 – 3/15/2010
ETC / EAC Revision (include Impact Analysis): _X_ Yes __No Corresponding ETC/EAC increase for additional software code development. Contl Acct. 123456.08.05.09.01.03 (WP03) ETC = +$78,000 Mar 2010; Contl Acct. EAC = $145,350 + $78,000 = $223,350	
Schedule Change: _X_ Yes __No Add a 6 week task to the CA 123456.08.05.09.01.03, (WP03). Impact is a 4 wk. schedule slip to start of Prototype SW I&T. Previous start to Prototype SW I&T was 2/18/2010, now start of Prototype SW I&T is 3/18/2010.	
Supporting Documents (Required): Attached, Yes / No or provide Link/URL _Y_ Cost Estimate _Y_ Schedule _Y_ Rationale & Basis of Estimate: 3 SW Developers x 240 hrs. x $108.33/hr = $78,000	
Approvals: Yes /No, Signature & Date Functional Manager: John R. Holmes Dec. 15, 2009 Subproject/Element Manager: Jimmy K. Stewart Dec. 15 2009 Project Manager: Sandra L. Smith Dec. 16, 2009	
Date Schedule Updated: 12/19/2009 Cobra Updated: 12/22/2009 WAD Updated: 12/23/2009 Comments: Project IMS has been updated to reflect 4 week slip to SW I&T. / Baseline Budget updated on 12/22/2009.	

Figure 8-1: Example Baseline Change Request for IMS

8.4 Re-planning

Internal re-planning is a term used to describe changes made to the time-phasing of remaining baseline work and budget plans within projects that have EVM requirements. Internal re-plans are kept within the authorized scope, schedule, and budget requirements of the project's PMB. Internal re-planning generally changes the shape of the project PMB to reflect more accurately the plan to complete the remaining work.

In-progress tasks are typically not re-planned unless there is significant work left to accomplish. In this scenario, it is recommended that the in-progress task is redefined such that the work accomplished thus far is the entire scope for that task and the task is then recorded as complete. Remaining work scope, with the corresponding budget and duration, is then transferred to a new task or tasks and re-planned.

Tasks that have not yet started may be re-planned (e.g., adjustments to duration and budget) as necessary within the project boundaries (e.g., scope, schedule, and budget requirements). Careful coordination within the project between the responsible technical, resource, and schedule personnel is necessary to preserve the integrity of the IMS.

8.5 Re-baselining

The incorporation of CCB-approved changes that have a significant impact on both internal and/or external baseline cost, schedule, and technical plans is called a re-baseline. Under circumstances where the project's existing baseline schedule is no longer achievable and measuring performance and tracking variances is meaningless, the baseline schedule can be re-baselined with the proper governance change board approvals. The result of project re-baselining provides a new Commitment and Management schedule baseline for the project.

Management decisions need to be made concerning the treatment of variances prior to the re-baseline activity. Such decisions are in regards to the retention or elimination of schedule variances, cost variances, or both for future planning reference, performance evaluations, or other management needs. Partial re-baselines to selected portions of the baseline schedule, can be implemented for the same reasons and using the same process as a full project re-baseline.

8.6 Current Schedule Control

The Standard industry practice for schedule management calls for the establishment of a "schedule baseline" to be used as a measuring stick for determining project accomplishment and performance. This same standard practice also calls for the establishment of a "current schedule" that reflects the current project model of all tasks and milestones, sequencing, durations, progress updates, codes, constraints, and resources. The current schedule is compared to the baseline schedule to measure project progress and performance. Analysis of the variances between the baseline and current schedule dates and durations provides necessary information for management decisions.

The control processes established for managing the current schedule should not be confused with the formal baseline change process already addressed in the previous paragraphs. The purpose of the current schedule is to provide an accurate reflection of what has been accomplished along with an accurate representation of how the future work will be carried out, regardless of what the baseline schedule shows. Maintaining this accurate data in the current schedule should not be hampered by formal control processes, but rather by informal rules. These informal rules may include, but are not limited to the following: close communication, coordination, and informal approvals from responsible technical leads, approvals by project managers, and continual assessment and analysis to ensure proper forecasting, progress, and changes are incorporated. The nature of this control can be standardized and tailored for the project, depending upon the size and complexity of the IMS and the needs of the project. Implementing IMS version control during the update process is another simple technique for ensuring that the project team and customers are using the latest schedule information. By incrementing the schedule version or release date, and keeping copies of prior versions, practical current schedule control can be maintained.

It should be understood that in carrying out the normal current schedule update, analysis, and maintenance processes, there may be issues and conflicts that are identified that will precipitate the need for a formal baseline BCR. In these situations a BCR will be initiated and processed in a manner as outlined in the previous sections.

Chapter 9: Schedule Reporting

9.1 Overview

Schedule reporting is the dissemination of meaningful information about the schedule's overall status, progress to date, and forecast to complete (see figure 9-1). Schedule reporting helps determine if the project's objectives are being met. The following sections within this chapter recommend various types of schedule reports and reporting guidance that will prove useful in managing a project or program.

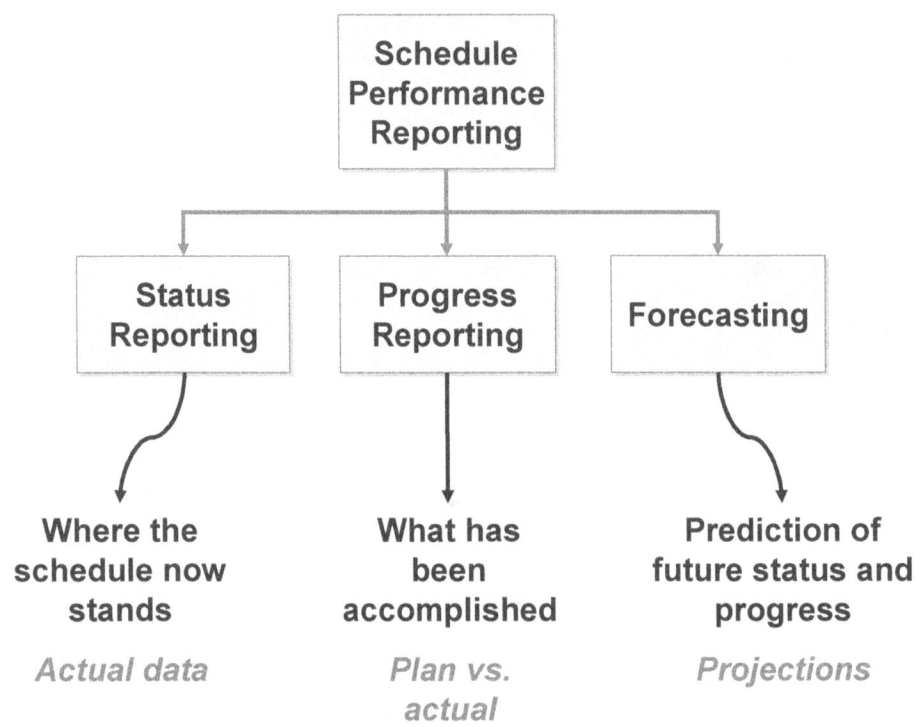

Figure 9-1: Schedule Performance Reporting

9.2 Best Practices

After a WBS is built, the reporting levels must be established. The basic assumption that all WBS elements must report at the same level is not a valid assumption. Schedule reporting will vary based on the level of management interest in the elements contained in the program or project. Higher volume (dollars or hours) or critical/risk items may require lower-level detailed reporting, while lower volume, non-critical/risk, or level-of-effort tasks may require only summary-level reporting. In today's programs/projects where resources are pushed to the limit, having flexibility in reporting requirements is a valid approach. The list below reflects some of the key considerations to keep in mind when categorizing element risk levels for reporting requirements (refer also to Figure 7-1).

A High Risk item ...

- Involves technological, manufacturing or other state-of-the-art advances or considerations.
- Is critical from the standpoint of achieving project objectives, reliability, maintainability, safety, quality assurance or other such factors.
- Is an item that program/project management designates as requiring special attention.
- May be pushing the amount of time allowed for development/procurement.

The paragraphs that follow provide sound guidance for recommended program/project schedule reporting. Caution should be taken to ensure schedule reporting requirements are consistent with overall schedule management requirements levied by NPR 7120.5, NPR 7120.7, and NPR 7120.8. While not specifically stated in these NPRs, it is important that all Gantt-type schedule reports contain a clear and visible legend of symbols used within each report. Due to the many different scheduling tools in use across NASA, and each having different symbol capabilities, it is not practical to require Agency-wide standardization. It is more practical for programs/projects to establish schedule symbology standards for use within their respective organizations. The real key is including a clear legend of symbols as part of each Gantt-type report to aid team members in accurately understanding the program/project schedules.

9.2.1 Management Summary

Each program/project should develop and make available a top level, Gantt-type summary document with program/project WBS elements identified. This document should, at a minimum, reflect all contract and controlled milestones, major development phases (i.e., design, fabrication, integration, assembly, etc.), clearly identifiable schedule margin, project critical paths, and all end item deliveries. This high level report is typically used by executive management to keep informed on general project progress and issues. This report does not provide enough schedule detail for thorough schedule analysis and is not intended for use as an analysis tool.

9.2.2 Schedule Logic Network

9.2.2.1 Program Schedule Logic Network

Each program should develop and make available an automated logic network database referred to as the Program Integrated Master Schedule (PIMS). This schedule report consists of schedule data for all project effort included in the program scope, along with all effort that only falls under the responsibility of the program organization to perform. As stated in Chapter 2 of this handbook (see section 2.1.1) the level of detail contained in the PIMS will generally depend on three key factors: the level of management insight desired by the program team, the magnitude of program scope and the amount of project data to be maintained and analyzed by the program team, and finally, the type and compatibility of schedule management tools used by all elements within the program. Due to these factors, the schedule content of the PIMS may be a compilation of all the project detailed schedules or, it could potentially be a compilation of summary versions of each project IMS. Schedule data for all effort that falls under the responsibility of the program organization should be reflected in lower discrete detail regardless of what level is included for the projects. This report can be used by the program team for identification and analysis of program/project issues and impacts, critical path analysis, "what-if" analysis, program schedule trending, and milestone accomplishment oversight.

9.2.2.2 Project Schedule Logic Network

Each project should develop and make available an automated logic network database referred to as the Integrated Master Schedule (IMS) consisting of schedule data for all WBS elements. The entire scope of work should be broken into schedule tasks and milestones at a consistent level of detail to allow discrete progress measurement and visibility into the overall design, fabrication, integration, assembly, test, and delivery phases of each end item deliverable. Additionally, all schedule tasks/milestones should be integrated with the appropriate sequence relationships to provide a total end-to-end logic network leading to each end-item delivery. The IMS should contain all contract and controlled milestones, key subcontractor milestones, end item delivery dates, key data delivery dates, and key Government Furnished Property (GFP) need dates. The database should contain the appropriate task coding attributes necessary to provide sort, select, and summarization capabilities for, but not limited to, WBS element, project phase, and level-of-effort tasks. The IMS serves as the basis for identification of project critical paths as well as critical detailed issue analysis, work-off and performance trending, schedule risk assessments, and "what-if" analysis.

Development and utilization of an IMS is an industry-wide "best practice" and is recommended for all types of NASA projects. An IMS is typically a contract requirement for all prime contractors, but is also strongly recommended for all project effort being accomplished by in-house NASA organizations.

9.2.3 Critical Path Identification

Each program/project should develop and make available a report for management use that identifies the critical schedule driver(s) for the program/project. This report should be an extract from the PIMS/IMS and include all tasks and milestones with a specified amount of total slack (float). It is recommended that all tasks/milestones with ten workdays or less of total slack be included in critical path reports. The report should be submitted in a waterfall format and organized in manner such that the path with the least amount of slack is delineated first and followed by each successive path according to total slack values.

9.2.4 Total Slack Report

Each program/project should develop and make available one or more total slack reports that show the history of slack for a particular area or group of program/project activities. This report may be used to show trends by WBS element, by milestone, or by any other logical grouping. The report should show the total slack for each reporting period. Reasons for total slack changes between reporting periods should be noted. An example of this type of report is contained in the chapter on schedule assessment and analysis (see Figure 7-7).

9.2.5 Schedule Risk

Each program/project should develop and make available a report indicating the probability of program/project completion within the approved timeframe. The frequency of such reporting will be event driven. Significant changes in technical risks, new emerging risks, retirement of risks previously identified, transition to a new project life cycle phase, baseline resets, and significant funding changes are all examples of events that may dictate the need for a schedule risk assessment. A variation of the aforementioned report may be used to identify the necessary timeframe, to a specified level of confidence, to complete an effort. An example of these types of reports can be found in the chapter on schedule assessment and analysis (see Figure 7-15).

Caution should be exercised when preparing a schedule risk assessment report at the program level. If the PIMS contains only summary-level project schedule data, then determining the necessary minimum and maximum duration parameters becomes much more difficult and can have extreme effects on the resulting schedule risk report data. This is due to the much longer summary-level task durations involved which can potentially cause a much wider range of risk impact in program durations resulting in risk probability data that may be invalid.

9.2.6 Schedule Margin Metrics

Each program/project should develop and make available a report clearly showing the trend for schedule margin usage over the life of the program/project. Any change between reporting periods should also be clearly explained. An example of this type of report can be found in the chapter on schedule assessment and analysis (see Figure 7-11).

9.2.7 Performance Trends

Each program/project should develop and make available a report clearly showing the trends relating to schedule performance. Such trends should include, but are not limited to accomplishments (planned finishes versus actual finishes), control milestone completions, forecasts (planned starts versus actual starts), and workloads (cumulative starts and finishes, planned versus actual). Significant changes between reporting periods should also be clearly explained. Examples of these types of reports can be found in the chapter on schedule assessment and analysis (see Figures 7-4, 7-5, and 7-6).

Chapter 10: Schedule Data and Lessons Learned Archival

10.1　Overview

Schedule data and lessons learned should be routinely archived in order to ensure important schedule information is not lost. Original baseline and as-built schedules, along with schedule versions at major project milestones should be the minimum for project archives. This information is essential for estimating, forecasting, and analysis efforts on current and future programs/projects.

10.2　Schedule Archives

Administrative closure is the process of documenting and archiving project data in a timely manner, as well as formally accepting the project's product and lessons learned. It includes verifying scope, archiving or maintaining project information, and producing summary information about items such as cost, work, and schedules.

Electronically archived schedules can provide a wealth of information. Care should be taken to properly label and store these datasets. Perhaps the two most important versions of the project schedule would be the original baseline and the "as-built" schedule. The original schedule baseline will provide future projects with a set of data to use as a reference point for initial planning of credible schedule content. The completed set of schedule data after all project work is completed is the "as-built" schedule, and represents what actually was implemented and how long each task took. This information is invaluable in future planning of task durations and determining schedule margin. Additionally, electronic schedule versions at key phase changes or events, or major baseline versions, also provide much useful information and should be archived.

As is the case with the lessons learned, archived schedules should be placed within a master database or commonly accessible repository.

10.3　Lessons Learned

"Lessons learned" is a review of best practices, project insights, and client information. As the project team disbands and new programs/projects begin, reviewing the lessons learned allows you to retain a record of the information gathered and generated through the project.

The end of a project is an opportunity to gather and record project information and share it with stakeholders.

A lessons-learned exercise helps in the collection of best practices and project data that can be shared through reports, white-paper views, the Internet, or other distributions. Historical schedule data can also be used for future programs/projects by saving schedule files as templates. Properly following through with these important last steps of a project closeout can positively affect the success of future projects.

A lessons-learned exercise documents best practices and how the project progressed, using information gathered from reports, discussions, or meetings. It includes information about a project's successes and failures, and can be used for a final team meeting, training on future projects, or recurring similar projects. Examining lessons learned provides an opportunity to answer questions like:

- Was the project objective completed?
- Was the work on time, within budget, and by specifications?

- What can we do to improve future projects?
- Were the stakeholders satisfied?

A lessons-learned exercise also provides information for administrative closure and contract closeout.

This process should address all phases of the project, from inception to completion. Careful attention should be given to tasks/activities that were very successful, as well as those that were not performed as planned. Effective techniques should be carefully recorded and replicated, where practical. Just as carefully, ineffective practices should be noted and avoided in the future.

Finally, there should be a process established to transfer this knowledge into a master database or repository. This facilitates the use of this data by others in the organization that could benefit from it. This master database may be maintained in business offices, systems management offices, or project analysis offices.

10.4 Historical Narrative

A historical narrative presents past project schedule information in a meaningful format with content defined and documented for the full project life cycle. When closing out a project, save reports to analyze task/milestone and resource information, start and finish dates, costs, and what actually happened versus what was scheduled to happen. It is very important that actual task durations are recorded to enable more meaningful usage of this data later. The following types of project reports should be archived:

- Schedule reports: Project Summary, Milestone, IMS at key project junctures
- Cost reports: Budget, Earned Value, Over-budget Tasks, Over-budget Resources
- Operational reports: Launch Readiness Review history, simulation testing, personnel certifications

Contract closeout ensures that the contractors' final scheduled work is completed and delivered and that billings or invoices are complete. At contract closeout, it's important to review schedules, changes, and contractor performance.

The historical narrative should present a story of the project from beginning to end that highlights the major events. Detailed data may be referred to, including direction regarding how to retrieve that data. The narrative should not capture every single detail, but provide others with enough information to determine how the project scope of work was accomplished and what implementation strategies were used.

10.5 Data Statistics

Schedule statistics are often useful to other P/Ss developing future projects. Most of these statistics can be gleaned from routine data backups and change control documents, if properly maintained. One approach in this area is to converse with others in the organization that are involved in schedule development, in order to ascertain the types of statistics that would be most useful for future development efforts. Preparation and dissemination of a summary document during closeout is often a much more efficient use of time than later researching archives.

Another good practice is to make a "wish list" during the project, recording the types of schedule-related statistics that you would have found helpful had they been available. During the course of the project, part of your effort should be directed toward, or at least cognizant of, accommodating this type of data collection at project completion.

Yet another good practice is to follow the practices mentioned elsewhere in this guide. Specifically, the use of change control, data structure, common naming conventions, and consistent data coding enhance the collection and analysis of data statistics. By maintaining structure and discipline in the scheduling function, the quality of the data that can be mined and clearly understood will be improved, and the data will be more easily and quickly acquired.

APPENDIX A: Acronyms

AA	Associate Administrator
ACWP	Actual Cost of Work Performed
ANSI	American National Standards Institute
APPL	Academy of Program and Project Leadership
ATP	Authority to Proceed
BCR	Baseline Change Request
BCWP	Budgeted Cost of Work Performed
BCWS	Budgeted Cost of Work Scheduled
BM	Business Manager
BOE	Basis of Estimate
BOM	Bill of Materials
BCR	Baseline Change Request
CA	Control Account
CAM	Control Account Manager
CCB	Configuration Change Board
CCR	Configuration Change Request
CD	Center Director
CDR	Critical Design Review
CDRL	Contract Data Requirements List
CO	Contracting Officer
COTR	Contracting Officer's Technical Representative
COTS	Commercial-Off-The-Shelf
CPI	Cost Performance Index
CPM	Critical Path Method
C/SRA	Cost and Schedule Risk Assessment
DCR	Design Certification Review
DR	Data Requirements
DRD	Data Requirements Document
EIA	Electronic Industries Alliance
EV	Earned Value
EVM	Earned Value Management
FAD	Formulation Authorization Document

FM	Financial Manager
FRR	Flight Readiness Review
GFP	Government Furnished Property
GFE	Government Furnished Equipment
GPMC	Governing Program Management Council
HOT	Hands-On Training
HQ	Headquarters
IMS	Integrated Master Schedule
IPT	Integrated Product/Project Team
IT	Information Technology
KPP	Key Performance Parameters
JCL	Joint Confidence Level
LCC	Life Cycle Cost
LOA	Letter of Agreement
LOE	Level of Effort
MOU	Memorandum of Understanding
MSFC	Marshall Space Flight Center
NASA	National Aeronautics and Space Administration
NOA	New Obligation Authority
NPD	NASA Policy Directive
NPR	NASA Procedural Requirements
OBS	Organization Breakdown Structure
OJT	On-the-Job Training
P/S	Planner/Scheduler
PCA	Project Commitment Agreement
PCM	Project Control Milestone
PDC	Probability Distribution Curve
PDR	Preliminary Design Review
PE	Principal Engineer
PEM	Project Engineering Manager
PIMS	Program Integrated Master Schedule
PM	Project Manager
PMB	Performance Measurement Baseline
PMBOK	Guide to Project Management Body of Knowledge

PMI	Project Management Institute
PMT	Performance Measurement Technique
POP	Project Operating Plan
PPBE	Program Planning Budget Execution
PRA	Project Resource Analyst
PRR	Preliminary Requirements Review
PSOBD	Project Schedule's Office Document
RBS	Resource Breakdown Structure
REDSTAR	Resource Data Storage and Retrieval
RFP	Request for Proposal
RM	Resource Manager
SATERN	System for Administration, Training, and Educational Resources for NASA
SCB	Schedule Change Board
SMH	NASA Schedule Management Handbook
SMP	Schedule Management Plan
SOLAR	Site for On-line Learning and Resources
SOW	Statement of Work
SRA	Schedule Risk Assessment
SRR	System Requirements Review
TA	Task Agreement
TF	Total Float
WAD	Work Authorization Document
WBS	Work Breakdown Structure
WP	Work Package

APPENDIX B: Glossary of Terms

- Actual Cost for Work Performed (ACWP) – The costs actually incurred and recorded in accomplishing the work performed within a given time period. (Actual costs include the direct cost plus the related indirect cost such as overhead, G&A, etc. allocated to the activity.)

- Baseline Schedule– the original approved plan plus or minus approved scope changes.

- Budgeted Cost for Work Performed (BCWP) (or Earned Value) – The sum of the budgets for completed work packages and completed portions of open work packages, plus the applicable portion of the budgets for level of effort and apportioned effort. This is the value of the work accomplished.

- Budgeted Cost for Work Scheduled (BCWS) (or Planned Value) – The sum of the budgets for all work packages, planning packages, etc., scheduled to be accomplished (including in-process work packages), plus the amount of level of effort and apportioned effort scheduled to be accomplished within a given time period. This is the value of planned work.

- Commercial Off The Shelf (COTS) – a term usually used to describe purchased software (as opposed to customized or internally developed software).

- Contracting Officer (CO) – a person with the authority to enter into, administer, and/or terminate contracts and make related determinations and findings Contracting Officer's Technical Representative (COTR) – person responsible to the CO for the contractor's performance on a given contract.

- Cost and Schedule Risk Assessment (C/SRA) - the process of performing a probabilistic risk assessment on a project schedule that is resource loaded. This type of probabilistic assessment is based on using monte carlo simulations that incorporate minimum, maximum, and most likely estimates for task durations.

- Cost Performance Index (CPI) – An indicator of the cost efficiency of the work accomplished for the current period(s) or cumulative-to-date as derived by the formula: CPI equals BCWP divided by ACWP, i.e., Earned Value divided by Actual Cost Incurred.

- Critical Path – A sequential path of tasks in a network schedule that represents the longest overall duration from "time now" through project completion. Any slippage of the tasks in the critical path will increase the project duration.

- Data Requirements (DR) – a set of datum that is mandatory for the purpose specified.

- Data Requirements Document (DRD) – the document containing the DR.

- Earned Value (EV) – The budgeted value of work accomplished. The value of completed work expressed in terms of the budget assigned to that work. It is the sum of budgets for completed work packages and completed portions of open work packages, plus the appropriate portion of the budgets for level of effort and apportioned effort. Also known as Budgeted Cost of Work Performed (BCWP).

- Estimate At Completion (EAC) – Actual direct costs, plus indirect costs allocable to the project/contract, plus the estimate of costs (direct and indirect) for authorized work remaining.

- Free Float (FF) – the amount of time a task may be delayed without impacting the start of any of its immediate successors.

- Guide to the Project Management Body of Knowledge (PMBOK Guide) - a publication authored by the Project Management Institute (PMI) that contains project management guidelines.

- Integrated Master Schedule (IMS) – An integrated schedule developed by logically networking all detailed program/project activities. The highest level schedule is the Master Schedule supported by Intermediate Level Schedules and by lowest level detail schedules.

- Key Performance Parameters (KPP) – quantitative metrics selected by the PM in order to measure the effectiveness of the project in achieving their goals and the related mission success criteria.

- Lesson Learned - the significant knowledge or understanding gained through past or current programs and projects that is documented and collected to benefit current and future programs and projects.

- Level of Effort (LOE) – Effort of a general or supportive nature that does not produce definite end products.

- Milestone - An event of particular significance. Finitely defined events that constitute the start or completion of a task or occurrence of an objective criterion for accomplishment. Milestones should be discretely identifiable, the passage of time alone is not sufficient to constitute a milestone. However, milestones should be associated with schedule data to document when the milestone is to occur.

- NASA Policy Directive (NPD) – HQ level document establishing NASA policy.

- NASA Procedural Requirements (NPR) – HQ level document that defines process requirements.

- National Aeronautics and Space Administration (NASA) – The U.S. government organization responsible for space and earth sciences, experimentation, and exploration.

- New Obligation Authority (NOA) – approval to obligate resources to the level specified.

- Organizational Breakdown Structure (OBS) – A family-tree breakdown of the performing activity's organization showing the organizational elements involved in performing the work.

- Performance Measurement Baseline (PMB) – The time-phased budget plan against which performance is measured. It is formed by the budgets assigned to scheduled control account and the applicable indirect budgets. For future effort, not planned to the control account level, the performance measurement baseline also includes budgets assigned to higher level WBS elements and undistributed budgets. It equals the total allocated budget less management reserve.

- Planner/Scheduler (P/S)- The person who performs planning and scheduling functions for a project or program. This person may be dedicated solely to this function or may share this function with other functions.

- Planning Packages (PP) – A logical aggregation of work within a control account, normally the far-term effort, that can be identified and budgeted in early baseline planning, but is not yet defined into work packages.

- Probability Distribution Curve (PDC) – a statistical representation of the likelihood of a value occurring during simulation that is between the minimum and maximum values.

- Program - A strategic investment by a Mission Directorate (or Mission Support Office) that has defined goals, objectives, architecture, funding level, and a management structure that supports one or more projects.

- Program Planning Budget Execution (PPBE) - a financial document that indicates planned expenditures or a funding request for future years for a project.

- Project - A specific investment identified in a Program Plan having defined goals, objectives, requirements, life-cycle cost, a beginning, and an end.

- Project Management Institute (PMI) – an international organization comprised of members of the project management community that share a common interest in promoting the development of the project management discipline.

- Request For Proposal (RFP) – a procurement term that describes the package of information given to a contractor or contractors that are invited to submit bid packages for a given scope of work.

- Resource Leveling – A process for the sequencing of schedule tasks/activities, without violating network logic, for a given resource or resources in a manner that results in a more consistent level of demand for that resource or resources over the life cycle of the project.

- Resource Loading – the process of recording resource requirements for a schedule task/activity or a group of tasks/activities.

- Schedule Logic Network – A schedule format in which tasks/activities and milestones are represented along with their relational interdependencies, constraints, and durations. It expresses the logic as to how the work scope will be accomplished. Logic Network schedules are the basis for critical path analysis, which is a method for identification and assessment of schedule priorities and impacts.

- Schedule Management - the establishment, monitoring, and maintenance of the baseline master schedule and derivative detailed schedules. It is composed of the establishment and operation of the system and includes (1) definition of format, content, symbology, and control processes, and (2) selection of key progress milestones and indices for measuring program and project performance and indicating problems.

- Statement Of Work (SOW) – a document that contains a narrative description of the work scope requirements for a project or contract.

- Schedule Performance Index (SPI) - An indicator of the schedule efficiency at which work has been performed to date. SPI equals BCWP divided by BCWS, i.e., Earned Value divided by Planned (Budgeted) Value.

- Schedule Risk Assessment (SRA) – the process of performing a probabilistic risk assessment on a project schedule. This type of schedule assessment is based on using monte carlo simulations that incorporate minimum, maximum, and most likely estimates for task durations.

- Slack (Float) - The terms, slack and float, are used interchangeably throughout this document and mean the same thing. Both refer to the amount of time that a task or group of tasks may be delayed without impacting the start of a later task or group of tasks. There are two types of float. Free float refers to the amount of time a task can slip before impacting the early start date of its immediate successor(s). Total float refers to the amount of time a task may slip before impacting project completion.

- Task (Activity) - In scheduling terms, a task or activity is the lowest level of detail shown in a schedule. It is a piece or portion of discrete, apportioned, or level-of-effort work. It represents effort that occurs over a period of time and generally consumes resources. For the purposes of their use in this document, these two terms should be considered synonymous.

- Task Agreement (TA) – a written document describing the relationship between entities and scope of work each is responsible for accomplishing.

- Total Float (TF) – the amount of time a task may be delayed without impacting the end date of a schedule.

- Work Breakdown Structure (WBS) - A product-oriented hierarchical division of the hardware, software, services, and other work tasks that organizes, displays, and defines the products to be developed and/or produced and relates the elements of the work to be accomplished to each other and the end products.

- Work Breakdown Structure Dictionary - A document that describes the tasks-associated with each WBS element, in product-oriented terms, and relates each element to the respective, progressively higher levels of the structure as well as to the Statement of Work.

APPENDIX C: Data Requirements Document (DRD)

DATA REQUIREMENTS DESCRIPTION (DRD)

1. DATA PROCUREMENT DOCUMENT NO.: XXX   ISSUE: Standard
2. DRD NO.:
3. DATA TYPE: 2
4. DATE REVISED:
5. PAGE: 1/3
6. TITLE: Project Integrated Master Schedule

7. DESCRIPTION/USE: To provide the contractor's time-phased plan, current status, key milestones, task interdependencies, and major development phases necessary to accomplish the total scope of work. This schedule will be used to provide management insight into contractor status, potential problem areas, and critical path identification, which will serve as the basis for evaluating contractor performance.

8. Office of Primary Responsibility:
9. Data Manager:

10. DISTRIBUTION: Per Contracting Officer's letter

11. INITIAL SUBMISSION: Preliminary with proposal. Initial - first calendar month following the end of the first full month after Authority to Proceed (ATP).

12. SUBMISSION FREQUENCY: Monthly, no later than the 10th day of the calendar month following the end of the contractor's calendar month.

13. REMARKS: The schedule will be baselined at some point after ATP as agreed to by both parties and not to exceed 90 days after ATP. Reference is made to NPR 7120.5 (Current Revision), *NASA Program and Project Management Processes and Requirements*. This document shall be used as a guide in preparation of the Project Schedules.

14. INTERRELATIONSHIP: DRD for *Work Breakdown Structure and Dictionary*, and DRD for *Project Management Plan*

15. DATA PREPARATION INFORMATION:

15.1 SCOPE: The Project Schedule provides data for the assessment of contract schedule and logic network of the tasks to be performed.

15.2 APPLICABLE DOCUMENTS:

NPR 7120.5 *NASA Space Flt Program & Project Management Requirements*
NPR 7120.7 *NASA Information Technology and Institutional Infrastructure Program and Project Requirements*
NPR 7120.8 *NASA Research & Technology Program & Project Mgmt Requirements*

15.3 CONTENTS: The project schedule shall include tasks necessary to accomplish the total scope of work as defined in the work breakdown structure (WBS). The schedule shall also include all logical relationships (interdependencies) between tasks. Schedules shall contain the approved baseline schedule as well as current forecasted dates and shall be traceable to the approved Work Breakdown Structure

(WBS). All key milestones shall be clearly identified and logically linked to related tasks. The project schedule shall be created and maintained in management software that supports automated time phasing of tasks, a logic driven critical path, schedule assessment, and trend analysis capabilities. Project Schedules and Logic Network shall be reported in four sections. The following deliverables shall be extractions from the automated logic network database. All data contained in the sections shall be consistent, statused monthly and based on the same cutoff date.

a. Summary Schedule – One page, top level, Gantt-type summary document arranged by WBS that reflects all contract and controlled milestones, major project phases (i.e., design, fabrication, integration, assembly, etc.) and all end item deliveries.

b. Logic Network Database – an automated logic network database consisting of schedule data for all WBS elements. The entire scope of work shall be broken into schedule tasks and milestones at a consistent level of detail to allow discrete progress measurement and visibility into the overall development, fabrication, integration, assembly, test, and delivery phase of each end item deliverable. Additionally, all schedule tasks/milestones shall be integrated with the appropriate sequence relationships to provide a total end-to-end logic network leading to each end item delivery. This database shall contain all contract and controlled milestones, key subcontractor milestones, end item delivery dates, key data delivery dates, and key Government Furnished Property (GFP) need dates. The database shall contain the appropriate task coding attributes necessary to provide sort, select, and summarization capabilities for, but not limited to, WBS element, project phase, and level-of-effort tasks. The logic network database serves as the basis for identification of project critical paths as well as critical schedule analysis.

c. Critical Path Report – This report shall be an extract from the Logic Network Database and include all tasks and milestones with 10 workdays or less of total slack (float). The report shall be submitted in a waterfall format and organized in manner such that the path with least amount of slack is delineated first and followed by each successive path according to total slack values.

d. Contractor Schedule Assessment Report – This report shall contain a count of the total number of tasks, milestones and non-detail (e.g., summary, hammock, rollup, etc.) activities contained in the schedule, a count of the number of completed tasks and milestones, a count of the number of tasks and milestones to be completed, a count of the number of tasks and milestones that have no predecessor and/or successor relationships, a count of the total number of tasks and milestones that have a total float (slack) value greater than 25% of the remaining duration of the total project schedule, a count of the total number of non-detail (e.g., summary, hammock, rollup, etc.) activities that have any predecessor or successor logical relationships, and a count of the total number of tasks and milestones that have forced or fixed dates. The report shall contain critical path narratives explaining changes and impacts to the critical paths listed in section c above. The report shall contain narrative explanations for contract milestones and significant project milestones that have moved more than 45 calendar days into the future from their baseline dates. Project milestones shall be identified and negotiated with the project office. These narratives shall include a proposed work-around schedule detailing how the contractor plans to recover the lost schedule time.

15.4 **FORMAT**: Submission of the deliverables in 15.3 shall be by standard hardcopy and electronic media. Electronic media submittals shall be in native file format utilizing schedule management software approved by the project office. A legend identifying the contractor's schedule symbols used and their meaning shall be provided.

15.5 **MAINTENANCE**: Changes shall be incorporated by change page or complete reissue.

Tailoring Note:

Special Tailoring Instructions from the OPR of this Standard DRD:

Special tailoring to this DRD should be incorporated only as-needed to meet unusual and/or specific project data needs and shall be coordinated with the responsible OPR.

Appendix D: – Schedule Training Topics

1. Work Breakdown Structure (WBS)
2. Scheduling Fundamentals
3. Critical Path Method (CPM) Scheduling Techniques
4. Logic Network Evaluation Techniques
5. Schedule Updates and Maintenance
6. Critical Path Identification and Analysis
7. Schedule Trend Analysis
8. Schedule Risk Identification
9. Probabilistic Schedule Risk Assessment
10. Schedule Resource Loading and Leveling
11. Schedule Analysis with Earned Value Data
12. Project Scheduling/Case Studies
13. Schedule Reporting (Guidelines and Formats)
14. Schedule Baseline Maintenance and Control
15. Structuring and Coding of Schedule Data
16. Managing Schedule Margin
17. Cost and Schedule Integration Techniques

Appendix E: – Schedule Management Reference Card

Page 1

Schedule Management Reference Card

Purpose of a Schedule: To provide a tool that supports the planning, directing, and controlling of a project in order to ensure its timely completion.

Scheduling Requirements: Key schedule requirements are found in NPR(s) 7120.5d, 7120.7 & 7120.8. These documents contain NASA's program and project management processes and requirements.

The Integrated Master Schedule (IMS) provides the time-phased plan for the total program/project work scope. It also serves as the foundation for cost & schedule integration and the Earned Value Management (EVM) baseline.

Benefits of an Integrated Master Schedule:

- Provides a roadmap for achieving project objectives
- Integrates project tasks into a logical sequence
- Identifies project critical path & forecasted completion date
- Establishes time-phasing for budget planning
- Provides a baseline to measure performance against
- Provides a capability to identify problems early & "what-if" analysis

Schedule Network Logic Elements:

- <u>Tasks & Milestones</u>: describes effort (keep discrete & measurable)
- <u>Relationships</u>: finish-to-start (preferred), start-to-start, finish-to-finish, start-to-finish (rarely used)
- <u>Durations</u>: length of task (use consistent time periods, measurable)
- <u>Constraints</u>: fixed dates (use sparingly & only when valid reason exists / overrides logic when calculating slack)

Page 2

Page 3

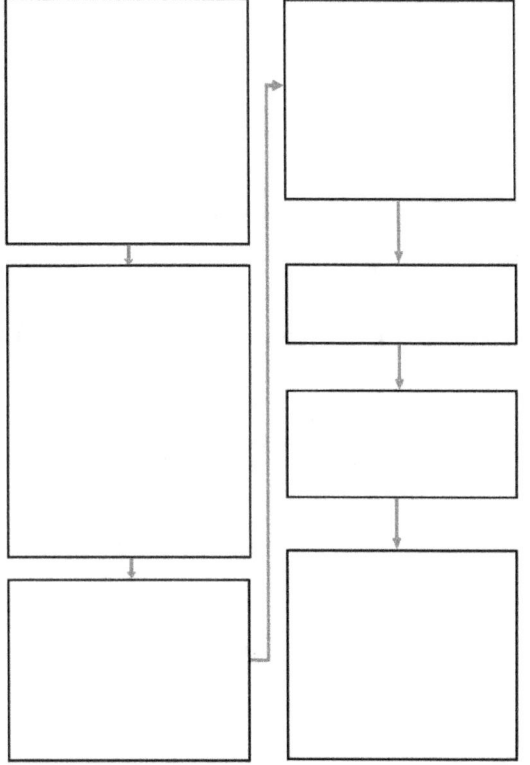

Page 4

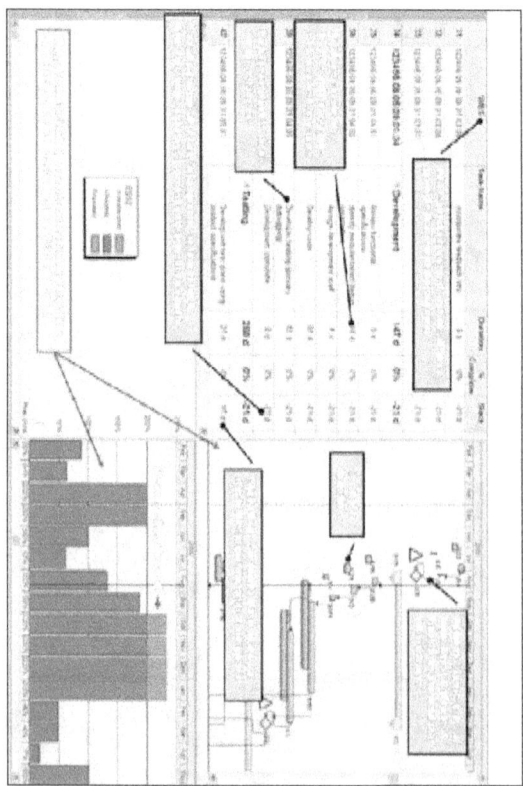

Appendix F: – Schedule Management Plan Template

Schedule Management Plan Outline

1. Introduction
1.1 Purpose
1.2 Reference Documents

2. Integrated Master Schedule (IMS) Approach
2.1 Schedule Automation
2.2 Master Logic Network
2.3 Baseline Management and Control
2.4 Contractor Schedules and Coordination
2.5 In-House Schedules and Coordination
2.6 External Organization(s) Schedules and Coordination

3. Project Schedule Management Organization
3.1 Roles and Responsibilities

4. Schedule Management Tool Considerations
4.1 Primary Scheduling Tools
4.2 Supporting Schedule Management and Analysis Tools

5.0 Project Integrated Master Schedule Development Processes
5.1 Work Packages
5.2 Planning Packages
5.3 Data Coding

6.0 Project Integrated Master Schedule Status Updates and Maintenance
6.1 Status Updates
6.2 Maintenance
6.3 Forecasting

7.0 Project Integrated Master Schedule Assessment and Analysis
7.1 Logic Network Integrity Assessment
7.2 Critical Path Analysis
7.3 Performance Trend Analysis
7.4 Schedule Risk Assessment

8.0 Project Integrated Master Schedule Baseline Control
8.1 Baseline Approval
8.2 Baseline Revisions
8.3 Baseline Re-plans

9.0 Project Integrated Master Schedule Reporting
9.1 Reporting Formats
9.2 Reporting Frequency

10.0 Schedule Data Archival & Lessons Learned Management

Appendix G: – Schedule Assessment Checklist

SCHEDULE ASSESSMENT CHECKLIST

#	Yes	No	Criterion Description
1.	___	___	Does the IMS reflect the total scope of work?
2.	___	___	Is the correct WBS element identified for each task and milestone in the IMS?
3.	___	___	Is the IMS used by all levels of management for project implementation and control?
4.	___	___	Do all tasks/milestones have interdependencies identified to reflect a credible logical sequence?
5.	___	___	Are task durations reasonable, measureable, and at appropriate level of detail for effective management?
6.	___	___	Does the IMS include all contract and/or designated management control milestones?
7.	___	___	Does IMS reflect accurate current status & credible start/finish forecasts for all to-go tasks and milestones?
8.	___	___	Has the IMS been resource loaded and are assigned resources reasonable and available?
9.	___	___	Is the critical path identifiable and determined by the calculated IMS logic network?
10.	___	___	Is the critical path credible?
11.	___	___	Has a Schedule Risk Assessment (SRA) been conducted on the IMS within the last three months?
12.	___	___	Has adequate schedule margin been included and clearly defined within the IMS?
13.	___	___	Has the IMS content been baselined and is it adequately controlled?
14.	___	___	Is there an excessive & invalid use of task constraints and relationship leads/lags?
15.	___	___	Are right task & resource calendars used in the IMS?

National Aeronautics and Space Administration
NASA Headquarters
Washington, D.C. 20546